Y0-ARS-711

DEATH OF A SCIENCE IN RUSSIA

Death of a Science in Russia

THE FATE OF GENETICS AS DESCRIBED IN *PRAVDA* AND ELSEWHERE

Edited by
CONWAY ZIRKLE
Professor of Botany
University of Pennsylvania

Philadelphia
UNIVERSITY OF PENNSYLVANIA PRESS
1949

Copyright 1949
UNIVERSITY OF PENNSYLVANIA PRESS
Manufactured in the United States of America

LONDON: GEOFFREY CUMBERLEGE
OXFORD UNIVERSITY PRESS

To that great company of Russian geneticists
and cytologists, now dispersed and destroyed,

to those who lost their positions and are denied
the exercise of their profession,

to those who simply disappeared,

to those who died under mysterious circumstances,

to those who, to save their families, recanted:

this book is respectfully dedicated.

Preface

At a time when the prestige of science is greater than ever before and when the benefits of scientific research are known to everyone, it seems incredible that a science should be destroyed in a country reputedly civilized and the scientists therein be forced to recant. If we consider further that the science destroyed was genetics at the very time it was making practical contributions to agriculture on an unprecedented scale, the event approaches the fantastic. When we note in addition that this occurred in a nation whose dominant ideology claims to foster science as one of its major objectives, we have obviously reached a point where we ought to find out what actually has happened. Unfortunately we have learned during the last few years that the mere fact that an event is preposterous does not mean that it cannot occur.

In its long and honorable career science has known many vicissitudes. Its practical benefits have nearly always been appreciated and, when it has been regarded popularly as just a part of technology, it has been highly esteemed, even though its methods and objectives were uniformly misunderstood. But science has always been more than a mere source of clever gadgets which add to our comfort. Scientists are continually trying to learn more and more about the universe and, as a by-product of this activity, they are continually discovering past mistakes. This aspect of their endeavors is not at all pleasing to many people, particularly to those who are emotionally attached to some endangered doctrine. In consequence, scientists have had to face the chronic hostility of the bigoted portion of mankind. Many times in the past science has been attacked and the scientists persecuted, but science has emerged unharmed from such ordeals and the scientists have been victorious ultimately. Science may be obliterated for a time in a particular region, but if it can remain free somewhere in the world its final triumph is assured. This is apparently admitted, for those who know its history hesitate to attack science. There are ignorant and backward communities, of course, where the *argumentum ad baculum* has been applied to scientists in the form of political action, but those who resort to coercing science have always been badly scorched, and organizations which have fought science in the past have shown themselves most anxious to forget any temporary successes they may have scored. Today the well-advised do not attack science.

To the world at large the Russian destruction of genetics was both inexplicable and shocking. It seemed so senseless in view of the great contribution being made by genetics to agriculture, due in great part to the activities of the Russians themselves. To many the event came without warning, but to the geneticists it was not entirely unexpected. There is a long record of hostility to genetics in Russia, a hostility which continued even when the Soviet Government was doing all in its power to breed better animals and plants. The hostility went underground during this period, but it emerged into the open a dozen years ago. Following the death of Lenin a number of high-ranking Communists encouraged the quackery which was to replace genetics, and in 1940 the charlatans scored their first real triumph. N. I. Vavilov, who was in charge of all research in agricultural genetics, was arrested and disappeared from the scientific world. Finally, in 1948, the blow fell and the final obliteration of genetics was begun in all the lands which owe allegiance to Stalin.

The winning clique acted with totalitarian thoroughness. Five geneticists found it expedient to recant, to discard their scientific knowledge, and to adopt the Communist orthodoxy. Nothing remotely like this has happened in the last three centuries. We can become indignant very easily at what has happened in Russia, and in fact we ought to become indignant, but if we are to understand what has happened and why, we need a certain amount of information.

It is the hope of the editor to make the pertinent facts available to the reader whether or not he is a trained geneticist. The information we have now not only enables us to learn how genetics was destroyed, but also helps us to discover how the Communist mind functions on the scientific level.

The editor has tried to keep his own writing at a minimum. In the introductory chapter he has furnished a brief historical background for the 1948 debacle and has tried to explain the issues which were raised, issues both real and pretended. Some of the perturbing happenings in Russian biology during the past decade already have been described and evaluated by several of our leading geneticists, who have most kindly permitted the editor to include their observations in this collection. These contributions will explain the forebodings of Western scientists and show why they were not taken completely by surprise when genetics was outlawed. All of this leads up to the final tragedy, the really damning facts of which will be presented in the words of the Russians themselves. No one who has tried to expose the Communists has ever succeeded in showing up the intellectual dishonesty of Party Line thinking as completely as have the Russians themselves.

Since Lysenko is the titular head of the forces attacking science, his presidential address to the Lenin Academy of Agriculture is printed in full. This is followed by the speeches of twenty-three of the other participants in the sessions. The speeches are published just as they were recorded in *Pravda*, as stenographic reports but shortened sometimes by the omission of unimportant or repetitive passages. The editor has supplied a number of explanatory footnotes to these speeches and, in some of the earlier ones, has pointed out a great many half-truths and complete falsehoods. However, it must not be taken for granted that the absence of a footnote means that the speaker was technically correct or personally honest. The same false statements were repeated over and over again in the same cant phrases and stereotyped expressions. To indicate every misstatement would be both tedious for the editor and boring for the reader. Also it is not at all necessary, as there are actually few real subtleties in the various Russian speeches. Most of Lysenko's followers are scientifically illiterate and are much too naïve to realize how their effusions would appear under a critical appraisal.

The climax of the tragedy is reached in the recantations of the scientists and in the abject and fawning letters written by the scientific academies to Stalin. When we read these degrading documents it would be well to remember that they were written in 1948 and addressed to the head of a great state; they were not written a thousand years ago and sent to the chieftain of some savage tribe.

The editor has been helped by a great many people whose assistance he wishes to acknowledge. He has been singularly honored by the wholehearted coöperation of his colleagues. First, Dr. Alexander Pogo, secretary and treasurer of the History of Science Society, read the articles in *Pravda* when they first appeared and insisted that the editor get them before the scientific public. His firmness made the editor undertake the task. The University of Pennsylvania, which aided the undertaking with a monetary grant, and the geneticists, who generously put their work at the disposal of the editor, made the project possible. The geneticists whose work is incorporated are, in order, Professor Richard Goldschmidt, Professor L. C. Dunn, Professor Karl Sax, Professor C. D. Darlington, Professor Th. Dobzhansky, Professor H. J. Muller, and Dr. S. C. Harland. Sir Henry Dale has kindly allowed the editor to reprint his letter of resignation from the Russian Academy of Science.

During the early stages of assembling the material for this book, news of the editor's plans spread among the geneticists, many of whom wrote him encouraging letters, offered assistance, and sent him news clippings and a great deal of material which he found most useful, even though much of it could not be included except in the bibliography. Every favor he asked of his colleagues was granted promptly, even at the cost of personal trouble and inconvenience. Those who have helped the editor locate important sources are Professor L. C. Dunn, Professor G. Ledyard Stebbins, Professor H. J. Muller, Professor C. D. Darlington, Professor John R. Baker, and Professor Th. Dobzhansky, who, even though he was in Brazil, sent valuable information and advice by air mail. The editor also wishes to acknowledge the help of Dr. G. H. Coons, Dr. Eubanks Carsner, Mr. Ben Dibner, Mr. Howard P. Barss, and Mr. Robert C. Cook.

Especial mention should be made of the editor's colleagues on the University of Pennsylvania faculty, Professor Robert E. Spiller, Professor George L. Haskins, and Professor Ralph Erickson, who read and commented on parts of the manuscript. Miss Marion Code, Miss Mary Ann Gehres, and the editor's wife, Mrs. Helen Kingsbury Zirkle, have helped in the editing and have striven to keep the work readable. The translators of the Russian articles, Mr. Stanley A. Kówalczyk, Dr. Beatrice Van Rosen, and Mrs. Jennie Appel, have shown a real interest in the work and, as a result of their special knowledge, have made many valuable suggestions. Finally, Mr. Phelps Soule, director of the University of Pennsylvania Press, should be mentioned especially as he has labored on the project over and beyond the call of duty. The editor wishes to express his gratitude to all of these.

The *Journal of Heredity,* the *American Review on the Soviet Union,* the *Nineteenth Century, Science, Hereditas,* and *Soviet Press Translations* have freely allowed the editor to include their copyrighted articles. Their kindness is acknowledged with thanks.

August 1, 1949 C. Z.

Contents

DEATH OF A SCIENCE IN RUSSIA

I

Introduction

1

OVERTURE TO TROUBLE

The first indication that all was not well with genetics in Russia was a news item in the *New York Times* of December 14, 1936. Prior to this, the lot of the Russian scientists had been very happy indeed. Even in the early days of the Revolution, days of disorganization and famine, when there were universal shortages and laboratory equipment was not to be found, special efforts were made to support and encourage science. Lenin, following the usual Marxian maxims, looked upon science as a major instrument for bettering the condition of humanity. The great Russian famine of 1920 had impressed him with the necessity of improving Russian agriculture, and he turned to genetics for help in securing improved varieties of domestic animals and plants. He placed enormous research facilities at the disposal of the great Russian geneticist N. I. Vavilov (1887-1942), under whose direction was made an enormous collection of useful plants, assembled from the entire world. Superior varieties were tested under Russian field conditions, valuable characteristics of different races were combined by hybridization, and new varieties were synthesized to order. A truly great group of investigators was assembled, given all the necessary equipment, and supported with seemingly unlimited resources as conditions improved. Russian genetics responded by making major contributions to the advancement of science. (See papers by Dunn, Darlington, Dobzhansky, etc.)

Some genetics laboratories were even established in the palaces of the displaced aristocrats, and the scientific workers were housed near-by in the smaller buildings on the great estates. All scientists had high social prestige in the Communist state, and here at last they found themselves esteemed and honored as they had always wanted to be esteemed and honored. Obstacles to research had been reduced to a minimum, and individual investigators were free to plan their own research within wide limits. Of course, in Russia, as in other countries, the scientists had a lunatic fringe, and the Russian fringe was decorated with some gaudy tassels, but this bothered no one. Then, too, popular ignorance and misconception of genetics were a little greater, perhaps, in Russia than in other countries, but these were all minor matters. Genetics was supported by the state, and the science and its devotees throve.

The news dispatch of December 14, 1936, came without warning and came as a shock. In fact, genetics had made such progress in Russia, and the geneticists of the world were so impressed with the accomplishments of their Russian colleagues, that they had planned to hold the 1937 International Congress of Genetics in Moscow. Preparations were well advanced and the hospitable Russians were getting ready for their guests when the whole project was suddenly halted. The dispatch created a sensation. It is quoted in part as follows:

MOSCOW CANCELS GENETICS PARLEY. . . . NAZI RACIST THEORIES ASCRIBED TO SOME SCIENTISTS CAUSE DROPPING OF WORLD CONGRESS. . . . AMERICANS WERE TO GO. . . . PROF. N. I. VAVILOFF, A FAMOUS PLANT EXPERT, IS ARRESTED . . . OTHERS UNDER ATTACK. . . .

Moscow, Dec. 13. The Seventh International Congress of Genetics, which was to have been held here next August, with a thousand of the world's leading scientists in this field participating, has been cancelled by order of the Soviet government, it is learned unofficially. Several British scientists who had expected to attend have been informed by Moscow that the congress will not be held. . . . About 100 Americans had been expected to attend. . . .

An interesting story of a schism among Soviet scientists, some of the most prominent among whom are accused by the Communist Party authorities of holding German Fascist views in genetics and even being shielders of "Trotskyists," lies behind the cancellation. The fact that so many of the Soviet Union's most distinguished geneticists are under fire is believed to be the motive for the government's action.

Attacks Classical Geneticists

In the past three months T. D. Lysenko, botanist, who has won great acclaim and high favor with the government for his experiments with the "vernalization" of wheat and other agricultural products to shorten the growing season, has been attacking the "classical geneticists" in the monthly scientific magazine, *Socialist Reconstruction of Agriculture.*

He challenged the validity of classical genetics, including the Mendelian laws and the chromosome theories, and stigmatized them as "formalistic" and of no practical value whereas his work, he said, is producing useful results. Mr. Lysenko said: "Genetics is merely an amusement, like chess or foot-ball" and he attacked the All-Union Institute of Plant Industry at Leningrad, headed by Academician N. I. Vavilov as useless.

The dispatch reported further that there were rumors of Professor Agol and Professor Vavilov being under arrest on charges involving Trotskyism, that the head of the Medico-Genetic Institution of Moscow, Professor S. G. Levit, was under fire charged with permitting in his institute "the development of scientific views hostile to Soviet theory and friendly to Nazis," an accusation which was surprising in view of his published work. Professor Shtyvko was attacked for an article on the skeletons of Russians who had died in the famine, in which he claimed they were intermediate between those of Germans and the yellow races. He had also said that the Buryat Mongols, a Siberian people, were the mental equals of twelve-year-old Europeans.

This news from Moscow caused a sensation, partly because of the rumored arrest of Vavilov, who was well known in America. Actually the rumor was false. It arose through a mistake (admitted by the *Times*) in the transmission of the dispatch. The sensation probably surprised the Russians, for on December 22 the *Times* carried an item which indicated that perhaps the Soviet authorities were hedging.

MOSCOW DEFENDS DELAY IN GENETICS . . . *IZVESTIA* SAYS TIME IS NEEDED TO COMPLETE PREPARATIONS FOR DELAYED CONGRESS . . .

Moscow, Dec. 21. The Soviet decision not to hold the geneticist congress scheduled for 1937 was confirmed today by *Izvestia,* chief government organ. It states the congress "has been postponed for some time at the request of a number of scientists who have expressed a wish to extend their preparation for the congress."

"Thus the only object of the postponement," *Izvestia* asserts, "is a desire to guarantee the best preparation and the utmost participation of scientists of various countries in its work."

No intimation is given when the "postponed" congress will be held.

Pravda denied the arrest of Vavilov, but admitted the arrest of Professor Agol for contact with "Trotskyist murderers." *Izvestia* attacked both the Medico-Genetics Institute of Professor Levit and the Soviet scientists who worked abroad and refused to return to Russia.

The next day, December 22, Vavilov delivered to the Moscow Bureau of the *Times* a statement which was also printed prominently in *Izvestia*. It was published by the *Times:*

VAVILOV DEFENDS SCIENCE IN SOVIET . . . CITES GROWTH OF STUDY OF BOTANY AND GENETICS OF PLANTS IN RUSSIA . . . PRAISES GOVERNMENT AID . . . HE SUGGESTS THAT ERRORS IN NEWS REPORTS ABOUT SOVIET ARE DESIGNED TO AID FASCISM.

Moscow, Dec. 22: "Your paper [writes Vavilov] published the totally false report of my arrest by the Soviet authorities. This is not the first time I have read false reports about myself in certain circles of the foreign press. Lies about Soviet Science and Soviet Scientists, honest workers for the cause of socialism, have become a specialty of the foreign press.

Sees an Aim to Discredit Soviet

"The publication by your paper of such falsification can only be explained by a desire by some circles to discredit the USSR and science in the USSR. Slander is the usual method of the enemies of the Soviet Union and is especially useful in furthering the cause of Fascism.

"Many times I have expressed myself in print and in lectures in many cities of the United States of America, about Soviet science, about the wonderful possibilities which are here available for the scientist, about the role of science in my country and about the great progress of science under the Soviets.

"From a small organization which existed in Czarist times, known as the Bureau of Applied Botany, the Institute of Plant Industry, which I have led since the coming the Soviets, has grown to a most important organization, having few equals in size among the institutes of the world.

"Its staff has increased from 65 people in Czarist times to 1,700 at present, including all of its subsidiary divisions. The budget has increased from 50,000 to 14,000,000 rubles.

"I can inform you and my colleagues that our work is making progress. We are developing important investigations in the selection and genetics of plants. The new building of the Institute of Genetics of the Academy of Science of the USSR is nearing completion, an additional evidence of the status of genetics in our country.

"We argue about and discuss the validity of the theories of genetics and the methods of selection. We challenge one another to socialistic competition, and I must tell you that this is a strong stimulus which significantly increases the results of our work. Our arguments occur chiefly in those fields where little has yet been decided by experiment or checked by practice.

Cites Discussion in Moscow

"Today at a special session of the Lenin Academy of Agricultural Science there is a discussion of problems of genetics and selection, and I am delivering an address on the subject of the paths of Soviet selection.

"I, more than many others, am indebted to the government of the USSR for the great attention it pays to the organization of which I am the leader and to my own work. As a true son of the USSR I consider it my duty and good fortune to work for the benefit of my fatherland and to give all of myself to science in the USSR.

"I want to say that the statement about me in your paper and the fabrications by Science Service indicating that in the USSR there is no intellectual freedom are totally untrue, and must come from hidden sources and I want this telegram to appear in your paper."

The pot continued to boil. On December 27 the *Times* published a dispatch which stated that *Pravda* had attacked G. M. Kaminsky, Commissar of Health, for protecting Professor Levit and the Medico-Genetics Institute, and for suppressing an attack on Levit by Professor Kroll. *Pravda* also stated that Levit had been expelled from the Party for making contact with counter-revolutionary elements and for being guilty of "a fascist scientific conception, a biological predestination of races, the omnipotence of heredity as the biological basis of criminality and so forth."

On December 28 another item in the *Times* described the discussion at the Lenin Academy of Agriculture to which Professor Vavilov had referred. Classical genetics was under heavy fire, and Lysenko of the Odessa Institute criticized Vavilov for his failure to take into account the experimental data on the production of new varieties of plants through intervariety crossing on fifteen thousand collective farms.

Professor Vavilov, on the other hand, said that the Odessa Institute had not obtained sufficient experimental data. Both men preserved all the amenities. Academician Zavadovski said that genetics had failed in the past to find a place in its classification for the work of the late Ivan Michurin and now for Professor Lysenko's work.

Professor H. J. Muller, the distinguished American geneticist, and later a Nobel Prize winner who was then working in Russia, denied the existence of a "formal" genetics, for even this early "formal" was being used as a meaningless derogatory adjective.

Pravda warned the All-Union Congress of Neuro-Pathologists and Psychiatrists to avoid theories which were current in Germany and which implied the unadaptability of physical workers for mental work.

All these attacks on genetics and geneticists ocurred toward the end of 1936. Early in 1937 the cold and patient war on genetics got under way. Vavilov in his telegram of December 22, 1936, was probably giving an accurate picture of genetics in Russia as it existed at the time, although we must remember that there was some evidence of hostility to geneticists. G. A. Leviski and N. P. Avdoulov were sent to labor camps in 1932, and at about the same time B. S. Chetverikov and W. P. Efroimson were sent to Siberia. Nothing was heard of them for fifteen years. (See page 74.)

Lysenko had started only a skirmish, however, and genetics was not to be hurt seriously for another three years. It would be allowed to linger on for an additional nine years. All the while the geneticists would be encouraged to hope that the Party line might change. Geneticists outside of Russia would also be given full opportunity for any wishful thinking in which they might indulge. It seemed incredible to them that in the twentieth century quackery could oust a science from a civilized country, and many of the American geneticists were waiting hopefully for Lysenko to be exposed. The Kremlin seemed to be undecided, for it continued to support both genetics and the Lysenko group. The contest on the surface had quieted down, but in 1940 the blow fell.

2

THE ISSUES, REAL AND BOGUS

Because today the destruction of genetics is being obscured by a great many irrelevant issues, we should pause for doctrine identification before we attempt to follow the march of events further. Not that doctrinal wrangling is the most important aspect of the Russian tragedy. It definitely is not. Of far greater significance is the evidence of the lack of intellectual freedom in a twentieth-century nation. When intellectual freedom does not exist, intellectual honesty becomes a liability, and the consequences for science are disastrous. These consequences are depicted without ambiguity in the Communists' own documents which are reprinted in the following chapters. We have proof of the destruction of science by dogma handed down from above (page 249), of the forced recantations of scientists whose knowledge is not acceptable to those in control (pages 265 ff.), and finally, of the punishment and liquidation of the scientists themselves (pages 288 ff.).

Attempts are being made to present the disturbance as a debate between two groups of scientists, as the sort of controversy which often occurs in a new field before sufficient data have accumulated to decide between rival preliminary interpretations. Of course, nothing of the sort is occurring, but this is not the major issue. Every thinking man should know just what the Russians have done to their science, and this is not a difficult task, for their acts are at last out in the open. They have published documents which contain innumerable records of their own political bigotry, records unequaled even by the Nazis, in fact unequaled anywhere during the past three centuries. The issue cannot be obscured or explained away. Attempts are made, however, to focus attention on certain technical aspects of biology, a field in which it is easy to confuse the nonprofessional. But even here the Communists have shown themselves to be extraordinarily clumsy. Anyone who wants to take the trouble can, with a very little technical assistance, recognize the bogus scientific standards in the official Russian "science."

It is, of course, impracticable to label every fallacy and misrepresentation in the published documents. They are far too numerous, and such a task would become tedious and repetitive. It is both amusing and instructive, however, to locate bits of odd logic in supposedly serious contributions, to find passages where disreputable motives are assigned to all who do not follow the Party line, and to locate other passages where denunciation is substituted for data. If we read these Russian documents understandingly, we shall learn much about the way the Communist mind functions.

We should remember that these articles were designed for the mass of literate Russians, and that the efficiency of their propaganda message depends both upon the partisan education of the readers and upon the fact that no printed refutation is allowed. The level of this propaganda is such that under more enlightened surroundings it should backfire. It is hoped that the following portion of this section will help to clarify the issue and thus aid the backfire.

RUSSIAN VERBALISM

This section is not a treatise on semantics. It has nothing to do with subtleties either of language or of meaning. In fact, it is little more than a guide to the more frequently used cant phrases, clichés, and stereotyped attitudes which serve either to establish the political orthodoxy of a Russian "scientist" or to attack the unconformity of an opponent. This use of language is highly formalized and, when encountered by someone unfamiliar with the Communist *Weltanschauung,* very confusing. As only a small part of the subject can be covered here, the reader is referred to Hudson and Richens[1] for a further account of the peculiar use of scientific terms by the Russians, even of those terms which are not directly involved in the Marxian philosophy.

The term "verbalism" is used here to denote not only an excessive attention to words but also a usage in which words do not label ideas but are substituted for ideas. In extreme cases it may even be synonymous with verbal amnesia, where the connection between words and their meaning is lost. Our first contact with this use of words is apt to be very puzzling, but it is necessary that we become familiar with it if we are to understand what the Russians say or write.

Many of us who followed the recent exchange of pleasantries between Marshal Tito of Jugoslavia and the Cominform doubtless felt vaguely baffled. The words exchanged were familiar individually, but they were combined into strange formulae, evidently of some meaning, for the two parties to the controversy seemed to understand each other. Obviously, this was an instance of a linguistic convention different from that to which we are accustomed. We should not dismiss such language, however, as mere Communist jargon. If we learn to understand it, we shall perhaps get a glimpse into the intellectual world in which the Communists live. To understand the Russian documents which are printed in this collection we shall have to allow for this odd use of terms and become familiar with at least three aspects of Russian verbalism.

First, we find that all Communist doctrines, theories, and hypotheses are personalized. Thus, the Russian universities do not have professors of evolution but professors of Darwinism. Genetics is rarely mentioned even in the heat of controversy, but instead such terms as "Mendelism," "Morganism," and "Michurinism" are current. This personalizing is much more than the mere attaching of personal labels to impersonal ideas. The ideas themselves are distorted. We can understand its implication better, perhaps, if we indulge in it once ourselves, just for the exercise. So, for the next paragraph, we will imitate the Russian method.

[1] P. S. Hudson and R. H. Richens, *The New Genetics in the Soviet Union.* Cambridge, England, Imperial Bureau of Plant Breeding and Genetics, 1946.

First, let us discard the term "law of gravitation" and use in its place a term derived from the name of its discoverer. We shall call it "Newtonism." Likewise, we shall discontinue the use of the word "relativity" in favor of "Einsteinism." But this is not all; following the Russian model we shall extend the terms. Newtonism will soon include all the ideas Newton ever had. It will not only mean the law of gravitation but also Newton's contributions to calculus, his mysticism, and in particular, his religion. Einsteinism will also grow until it includes Einstein's views on violin playing, his religion, and his views on international coöperation. We are now in a position to manipulate Newtonism and Einsteinism as expediency demands. We can even show that the two doctrines are incompatible without mentioning either gravitation or relativity. If in addition one of the terms, say Newtonism, had been endorsed by an authority so lofty that for one to express a doubt of Newton's infallibility would be to court destruction, we could outlaw Einsteinism without making the intellectual effort necessary for understanding relativity. This is exactly parallel to what the Russians have done when they placed Darwinism in opposition to Mendelism as a preliminary to the extirpation of the latter. The value of personalized terms for such an act is obvious, for even the Central Committee has not yet accused the geneticists of not believing in evolution.

The second function of Russian verbalism is to obfuscate the issues at crucial points by means of playing a sort of word-game. The object is to separate the words from their meanings. The usual method is to make the word contradict itself within a single sentence or in adjacent sentences. If this is done violently enough or skillfully enough, it ensures that from there on the argument can be followed only by ear. Two examples of this practice will be cited. The first is taken from an article by M. B. Mitin (reprinted here on page 42):

> However, we must be much more cautious than to take such a definition of gene as "materialism." We must remember that there is a mechanistic, vulgar, *metaphysical* materialism irreconcilable with the theory of evolution, and there is dialectical materialism, founded on the theory of evolution.

"Materialism" is thus taken out of circulation as a scientific term and becomes only a word to conjure with. The second example is taken from the words of Lysenko himself in the article reproduced on page 253:

> Is the role of the chromosomes belittled by what has been said? Not in the least. Is heredity transmitted through chromosomes during the sexual process? Of course, how could it be otherwise!
>
> We recognize chromosomes and do not deny their presence. But we do

not recognize the chromosome *theory* of heredity. We do not recognize Mendelism-Morganism.

"Chromosomes" are thus put in their place, but the place itself defies discovery. If the reader finds no sense in Lysenko's words, there is no need for him to assume that he is beyond his depth in an abstruse scientific subject. Lysenko's words do not make sense and are not intended to make sense. The reader might be amused in searching out the numerous examples of this word-game in the papers translated from *Pravda.*

An interesting comparison of the intellectual standards of the Nazis and the Communists suggests itself at this point. The Nazis were once exhorted by their leaders "to think with their blood" and, judging from the consequences, perhaps they tried to do so. The Russian readers who have access to only the Russian press are also set a novel anatomical task. The issues are presented to them in such a way that they are allowed to think only with their ears.

In considering the third aspect of this verbalism, we must realize that the crucial words have a definite rank, they are actually in a hierarchy. A few examples chosen at random should help to make this clear. "Dialectical materialism" of course is at the top. "Materialism" by itself is very good indeed, but certain adjectives may degrade it. Thus "bourgeois materialism" has sunk so low that it is an invective. "Idealism" is always bad because it is antithetical to materialism. Any term is also bad when preceded by "metaphysical" or "capitalistic." "Formalism" is damning, and "mathematical" is derogatory when applied to a biological problem. "Reactionary" means almost the same in Russia as it does in the United States, but in Russia it is applied to more subjects. In neither place, however, is it as bad as "Fascist beast" or "Trotskyist murderer." "Practical" is very complimentary when applied to a science or a scientist. "Erroneous" and "mistaken" mean just what we would suppose, except that they are nearly always used without qualification and are practically never accompanied with any citation of data. Anything is "erroneous" when it does not fit into an accepted theory. "Inadmissible" is the term used to dismiss an idea with the greatest possible speed.

One final aspect of this subject should be noted. It is that, when words are meaningless, they sometimes become very shifty and alter their rank in the hierarchy with great speed. These changes may be either positive or negative. Terms once derogatory may achieve eminent respectability, or the reverse may happen. Thus "Lamarckism," a damning label in 1940 (page 46), had acquired some Michurinistic prestige in 1948 (page 156). On the other hand, "Weismannism," relatively obscure in the period 1936-40, has become a major symbol of reaction in 1948. In fact, since last September it has tended to replace "Mendelism." "Mendelism-Morganism," which received great condemnation in July and August, is used only rarely in later issues of *Pravda,* its place being taken by "Morganism-Weismannism." Could this be due to the great pride Czechoslovakia has taken in Mendel? Will his name be saved even if the science for which he did so much is destroyed?

Enough examples have been cited to call the attention of the reader to the unusual use of language in the *Pravda* articles and, perhaps, to help him recognize other odd conventions.

ORGANIC EVOLUTION

Organic evolution is not directly involved in the controversy which destroyed genetics. It is referred to at times but always incidentally, and generally in such a way as to misrepresent the attitude of someone under attack. There is really no exact Russian equivalent for the term as it is used in the rest of the world. Darwinism, the Russian word, is not synonymous with evolution. A brief sketch of the theory is included here, however, for two reasons. First, most of the doctrines discussed in the controversial articles have played a major role in the development of our understanding of evolution, and their relationship to each other and their interactions can be presented most economically by a short historical account. Second, the individuals who have given their names to the controversial issues have all attempted to solve one or more of the problems involved in explaining what caused evolution to occur. Lamarckism and Darwinism will be mentioned briefly in their proper places, but they have become so prominent in Russia that they will have to be treated individually.

Evolution in the modern sense really belongs to the nineteenth and twentieth centuries. It is a complex of many hypotheses and theories, some of which are very old. Any adequate historical account of evolution should begin with the earliest written records. Here, however, we will be concerned only with those aspects which are necessary for understanding the action of the Russians.

Classical and medieval writers did not believe that species were constant, unchanging units which always bred true to type. Theophrastos described the alteration of plants by cultivation and the changing of one species into another by transplanting it to a different country. Virgil, in his Georgics, and Pliny told how the poor farmers planted wheat and barley only to find that the seed suddenly produced wild oats. Lysenko reminds us of this when, in 1948, he states that 28-chromosome wheat suddenly became 42-chromosome wheat (page 132). Of course, neither the Roman farmers nor Lysenko used pure lines of clean, pedigreed seed. St. Augustine (420 A.D.) described how seedling olives "degenerated," and Peter of Crescenzi (1305) and Levinus Lemnius (1561) wrote copiously on the transmutation of species. It was not until the middle of the eighteenth century that systematic investigation showed species to be relatively stable units. Another hundred years were to pass before biologists were convinced that species really evolved, but that the time element in evolution was of a different order of magnitude from what the earlier naturalists had supposed.

Lamarck, at the beginning of the nineteenth century, had assumed that acquired characters were inherited and that the process was cumulative from generation to generation, producing in time new species. In 1859 Charles Darwin both proved that evolution had occurred and explained its occurrence by natural selection.

Now natural selection, as understood by Darwin, was not a complete explanation of evolution. It would not account for the deterioration of an organ or explain how an organ could evolve through a useless phase in the process of acquiring a new function. Natural selection could not explain how cave fish became blind or how the penguin's wing evolved from a good wing for flying to a poor wing for flying, to a poor wing for swimming, to a good wing for swimming. The inheritance of acquired characters, however, was an excellent supplementary explanation, and Darwin, who earlier had spoken of "Lamarck's nonsense," adopted the latter's views and in 1868 tried to find a mechanism for the transmission of acquired characters.

By 1868 evolution was both proven as a fact and adequately, if not accurately, explained. The next step was to check the explanations by laboratory experiments. Difficulties immediately arose. Innumerable experiments designed to prove the inheritance of acquired characters gave negative results. A few, where positive results were reported, gave negative results when repeated. Many of these experiments had been so badly designed that their results were ambiguous and showed no traces of the inheritance of acquired characters when they were properly controlled. Then in the 1880's August Weismann appeared on the scene. Since he has become one of the supervillains in Russia, where he is described as a tool of reactionary capitalism, we should examine what he really did.

Weismann's contribution is designated by the phrase "the continuity of the germ plasm." He showed that in an early embryonic stage, before the cells are differentiated, a few cells are set aside and do not enter into the formation of the embryo proper. These are the germ cells which ultimately produce the eggs and sperm that unite to form the next generation. Each generation is thus produced not from the soma (bodies) of the preceding generation but from the germ plasm, which is continuous from generation to generation and is potentially immortal.

This work of Weismann made the inheritance of acquired characters extremely improbable, and definitely put the burden of proof upon those who accepted the hypothesis. We can illustrate its implications by an extreme instance. Worker bees are highly evolved and specialized. They are sterile females. For them to transmit the characteristics they acquire during their active and industrious lives would involve a minor miracle. It would be like girls inheriting the characteristics acquired by their maiden aunts. If we are produced, not from the bodies of our parents but from the never-ending stream of germ plasm, the difficulties of inheriting acquired characters are manifest.

Laboratory studies of natural selection also led to complications, and the first attempts to measure its effectiveness accurately were disconcerting. The villain this time was Wilhelm Johannsen, and he also is being denounced in Russia. At the beginning of the twentieth century Johannsen attempted to measure the effects of selection on the size of beans. A single generation of selection produced results. The progeny of the larger beans were larger than the progeny of the smaller beans. The difficulties arose in the subsequent generations. Continued selection produced no changes whatever. Beans in each selected line produced offspring of the same size regardless of their own size. To understand Johannsen's results we must remember that beans are normally self-fertilized. A population of beans thus consists of many inbred *pure lines*. Johannsen's positive results in his first selection came from separating the pure lines. His lack of further progress was due to the fact that *within* a pure line the variations are due to environmental factors and are not inherited. Unfortunately these variations were mislabeled "Darwinian variations."

Hugo De Vries had also been trying to explain evolution. He had observed large, sudden, inheritable changes occurring in stocks of *Oenothera,* the evening primrose. He named these changes "mutations," a term used today in a different sense. It might be better if we used the term "De Vriesian mutations" as synonymous with those larger changes involving chromosome alterations, and equate Darwinian variations with what geneticists now call "point mutations."

At any rate, at the beginning of the twentieth century evolution was an accepted fact, but its explanation was not in a happy state. It became obvious that the problem of evolution was a problem of heredity, for evolution could come about only through changes in heredity. The attention focused on heredity led to the discovery of Mendel's forgotten paper, and in it, and in the rapidly developing science of genetics, the clues to the causes of evolution were found.

Unfortunately, evolution has become extremely technical. Tremendous progress has been made, and problems which baffled the nineteenth-century biologists have been solved, but no one has succeeded in making the really complex factors understandable to the general public. It takes college students about one year past the elementary level to cover the ground. So much has been added to our knowledge of evolution in the last fifteen years that earlier work is now obsolescent. During the last few years some truly excellent books on the subject have appeared. Those listed in the footnote are invaluable to anyone who might wish to pursue the matter further.[2]

The present status of evolution theory can be given here only briefly and inadequately. The inheritance of acquired characters has been definitely discarded. Natural selection is still considered a major factor. Although new forms may arise without the assistance of natural selection and, in exceptional circumstances, in opposition to it, natural selection always remains a court of last resort and decides finally whether the new form will persist or perish. Mutation pressure, a force unknown to Darwin, is recognized as a major factor in evolution, and is beginning to be evaluated quantitatively. A third factor, the scattering of variability resulting from the chance combination of genes, becomes of real importance in small populations or breeding groups. The interaction of these three factors accounts for evolutionary change. Under various circumstances one or the other of them may play the dominant role.

INHERITANCE OF ACQUIRED CHARACTERS

In spite of the scientific advances of the last seventy years, the Communist Party has decreed that the inheritance of acquired characters must now be accepted by Russian biologists. It is interesting to note, in view of the Russian reverence for words, that the phrase itself has only recently been brought out into the open.

[2] Theodosius Dobzhansky, *Genetics and the Origin of Species.* New York, 1941. Ernst Mayr, *Systematics and the Origin of Species.* New York, 1942. Julian Huxley, *Evolution, the Modern Synthesis.* New York and London, 1942. George Gaylord Simpson, *Tempo and Mode in Evolution.* New York, 1944.

Although on the popular lower level the doctrine has been encouraged in Russia since 1920, even as late as 1944 Lysenko avoided it in his *Heredity and Its Variability*. Actual claims of the occurrence of such heredity were expressed in other words. In the Western world the name most generally attached to the doctrine is Lamarck's, but this name, like the doctrine itself, has been in general disrepute for about half a century. It will be interesting for us to learn whether Lamarck himself will be rehabilitated in Russia. If the Great Men of Russia are wise and well informed he will not be, as he is too easy to ridicule.[3] Besides, as his doctrine is an integral part of Michurinism, a most reputable, orthodox, and "safe" faith in Russia, there is no real necessity for resuscitating him.

Belief in the inheritance of acquired characters is very old. It can be traced back to the earliest myths, and we have some very plausible guesses as to how it originated. It was only necessary for the Stone Age agriculturist to combine, somewhat naïvely, two easily made observations: first, that domestic animals and plants are altered by environmental changes; and second, that children do, in general, resemble their parents. The Neolithic farmer, unconscious of the selection which was going on in his crops and herds, could hardly come to any other conclusion. The myth of Phaëthon shows how old the belief is. Phaëthon, the half-human son of Apollo, inveigled his father into allowing him to drive the chariot of the sun across the sky. The horses who drew the chariot ran away and carried the sun so close to the land of the Ethiopians that the inhabitants were scorched black. This character was passed on to their children, and today the Ethiopians are still black.

Belief in the inheritance of acquired characters has been carried along on the folklore level to the present day. Excellent examples are to be found in the *Just-So Stories* of Rudyard Kipling, where the elephants' trunks are explained by the fact that once, long ago, an elephant had his nose stretched out by a crocodile. Also, the camels acquired their humps through a similar appropriate experience, and animals in general have obtained their distinctive char-

[3] The inheritance of acquired characters was accepted by nearly all biologists when Lamarck was writing. His treatment of the subject was so absurd, however, that he made it ridiculous and suspect. He is responsible for the following gems: Cranes did not like to get their bodies wet when wading, so they stretched their legs and then found that they had to stretch their necks to reach the water where they feed. Thus they got long legs and long necks. Geese did not mind getting their bodies wet, so they lengthened only their necks. Giraffes became tall through stretching both their necks and forelegs. Ruminants obtained horns through a frequent loss of temper and a habit of butting. Their blood ran to their heads and deposited the bony matter which accumulated through the generations, etc.

acteristics through interesting adventures which make good stories.

The belief in inheritance of acquired characters is recorded also in classical and medieval philosophy. It was endorsed by Hippocrates (c.400 B.C.), Aristotle (384-321), Antigonus (285-247), Strabo (7 B.C.), Galen (130-220 A.D.), Solinus (235-300), and Justinus (400). It was accepted universally throughout the Middle Ages, indeed it was not until after Lamarck used it to explain evolution that any real skepticism developed. As the editor[4] recently traced the early history of the belief, there is no need to repeat the details here. Nearly ninety endorsements of the doctrine were recorded before Charles Darwin's, but only about half a dozen expressions of disbelief were found.

The inheritance of acquired characters lost its scientific status during the last quarter of the nineteenth century. Experiments designed to prove it gave negative results, and it became increasingly incompatible with the growing biological knowledge. Today it is about on a level with the Ptolemaic solar system.[5]

Belief in the inheritance of acquired characters did not, however, die everywhere at once. The editor has known a number of biologists who remained faithful to the doctrine, but they were all born before 1870. After a recorded life of 2,300 years, belief in such inheritance would hardly disappear overnight. Indeed, one of its faithful adherents visited Russia, received recognition and high honors, and was slated for a most important post. His suicide possibly disrupted a Communist time schedule and may have extended the life of genetics in Russia by many years.

THE KAMMERER CASE

This man was Paul Kammerer. During the first quarter of the twentieth century Kammerer, a Viennese zoölogist, published a

[4] Conway Zirkle, "The Early History of the Idea of the Inheritance of Acquired Characters and of Pangenesis," *Trans. Amer. Phil. Soc.* 35:91-151. 1946.

[5] Attempts are being made to confuse the issue in the Party line press by some of the American reviewers of Lysenko's book. First, the inheritance in unicellular organisms known as *dauermodification* is being cited as an instance of the inheritance of acquired characters. Such induced changes have been known for some years. Here, where there is no separation of germ plasm and the body, where the whole organism is both germ and soma, induced modifications are heritable, at least for a time. Geneticists have long been familiar with the inheritance of modifications induced in the germ plasm (mutations). We should expect such changes in unicellular organisms due either to alterations in the genes or in the cytoplasmic inclusions to be inherited, but this has no relevance to the disputed issue.

Second, the problem of cellular differentiation during embryonic growth is confused with inheritance in undifferentiated germ cells.

number of papers summarized later in a book[6] in which he claimed to have proved the inheritance of acquired characters. He based his case upon two series of experiments. He claimed first that the black, viviparous Alpine salamander and the black and spotted oviparous salamander of the lowlands could each be made to acquire the characteristics of the other. By breeding the black one on a yellow background and the spotted one on a dark background, he claimed that he could make each of them assume the appearance and breeding habits of the other and that these transformations became hereditary.

The second experiment was done with the midwife toad, the male of which grabs the egg strings laid by the female, winds them around his hind legs and carries them around until hatching time. As an exception to the rule, this male is devoid of the pigmented thick thumb pads which other frog and toad males need for the standard type of copulation. Kammerer claimed that he could change the mating habits of the midwife toad into those of ordinary toads by breeding them under the conditions of other toads. Such males not only stopped their "midwifery" but in addition developed the thumb pad and, again, this new induced character was said to be inherited and to mendelize if crossed to the original form. In the discussion of these claims statements were found in Kammerer's papers which did not tally. There was not enough time, according to his own records, for the generations he claimed to have bred.[7]

These claims of Kammerer received a great deal of publicity, and geneticists hastened to examine and test them. Proof of hereditary acquired characters in salamanders could not be established by preserved specimens. It would only be necessary to switch the labels of two museum jars to fake the evidence. The case was different, however, with the midwife toad. If this toad inherited nuptial thumb pads, preserved specimens should show it. Kammerer published photographs of what he claimed were black nuptial pads, but the photographs showed nothing clearly. Finally in 1919 William Bateson challenged Kammerer to submit his evidence, now reduced to a single preserved specimen, to examination. A most bitter correspondence ensued.[8] After refusing to allow Bateson to examine the thumb pads critically, Kammerer continued to claim that "dozens of scientific men had seen the pads and been convinced." The pressure on Kammerer was mounting, however, and in 1926 G. K. Noble, of the American

[6] Paul Kammerer, *The Inheritance of Acquired Characteristics*. New York, 1924.

[7] Quoted from Richard Goldschmidt, Presidential Address to Phi Sigma Societies. Professor Goldschmidt kindly allowed the editor to quote from his manuscript.

[8] See *Nature* 103:344-45. 1919. Also *Nature* 111:738-39. 1923, and 112:391, 899. 1923.

Museum of Natural History, and Hans Przibram, Director of the Institute where Kammerer worked, examined the specimen. They reported their sensational findings in adjacent papers in *Nature* (118:209-11. 1926). They stated that the so-called pads lacked the characteristic cells of the pads of other frogs and that their black pigment was injected India ink. Kammerer wrote the following letter, here reproduced from *Science* (64:593-94. 1926), and killed himself.

PAUL KAMMERER'S LETTER TO THE MOSCOW ACADEMY

This letter was sent to the officials of the Moscow Academy of Science by Dr. Paul Kammerer, Professor of Biology in Moscow University, a few days before his death.

Vienna, September 22, 1926.

To the Presidium of the Communist Academy, Moscow.

Respected Comrades and Colleagues:

Presumably you all know about the attack upon me made by Dr. Noble in *Nature*, of August 7, 1926. The attack is based upon an investigation of the exhibits of alytes (toads) with heat stripes, proving my theory, made by Dr. Noble with Professor Przibram in the Vienna Biological Experimental Institute and with my permission.

The principal matter of importance in this is an artificial coloring, probably with India ink, through which the black coloring of the skin in the region carrying the stripes is said to have been faked. Therefore it would be a matter of deception that presumably will be laid to me only.

After having read the attack I went to the Biological Experimental Institute for the purpose of looking over the object in question. I found the statements of Dr. Noble completely verified. Indeed there were still other objects (blackened salamanders) upon which my results had plainly been "improved" post mortem with India ink. Who besides myself had any interest in perpetrating such falsifications can only be very dimly suspected. But it is certain that practically my whole life's work is placed in doubt by it.

On the basis of this state of affairs I dare not, although I myself have no part in these falsifications of my proof specimens, any longer consider myself the proper man to accept your call. I see that I am also not in a position to endure this wrecking of my life's work, and I hope that I shall gather together enough courage and strength to put an end to my wrecked life tomorrow.

I am not stopping the packing up of the things destined to be taken with me. First, because it would attract the attention of my family, which must not know anything of my intention before it is carried out; and second, because I am thus making my last will and testament giving my library into the care of the Communist Academy at Moscow, so that this will compensate it for all the efforts it has wasted upon me.

Finally, I ask that my heartiest farewell greetings be given to the following friends: . . .

With the plea that you will forgive me for having made you all this trouble, I am,

Yours devotedly,
PAUL KAMMERER.

American biologists as a whole have tended to excuse Kammerer *(de mortuis nil nisi bonum)* and blame some overzealous assistant who wanted to give the master what the master so obviously wanted. Such things have been known to happen. Before we can accept this charitable interpretation, however, we shall have to explain Kammerer's seven-year reluctance to have his specimen examined, and his prolonged and skillful evasions of his critics' demands.

There is a distinct possibility that Kammerer will be resurrected as a Socialist hero. The University of Moscow, which has a museum containing the busts of the great evolutionists, includes Kammerer's in the collection. Foreign geneticists who visited Russia in the year following Kammerer's death have reported that some of their guides and translators thought that Kammerer was a persecuted and martyred friend of the workers. This popular attitude can be explained, perhaps, by the fact that a Soviet moving picture was based on Kammerer's life. Professor Richard Goldschmidt saw the film in 1929 and has kindly allowed the editor to quote the following from his unpublished manuscript:

One day, walking along the streets with my friend Philipchenko, I read in front of a movie house a huge sign of "Salamandra" decorated with pictures of this harmless animal. My surprised question was answered by my friend with the offer to see the film. This we did and my friend interpreted the text.

The film "Salamandra" turned out to be nothing but a propaganda film for the doctrine of the inheritance of acquired characters. It uses the tragic figure of Kammerer, his salamanders, and mixed up with them his midwife toads for the story. The importance attached to the subject is revealed by the fact that nobody else but the then all-powerful commissar for education, Lunacharsky, is the author of the film, that his wife plays the leading lady and that Lunacharsky himself appears on the screen in one scene, playing himself. Leaving out the interwoven love story written to fit the beautiful Mme Lunacharsky, the plot is this: In a central European University a young biologist (model Kammerer) is working. He is a great friend of the people and endowed with all the qualities of a communist movie hero. He works with salamanders and has succeeded in changing their color by action of the environment, and one day the supreme glory is achieved, the effect is inherited. The bad man of the play, a priest, has learned of this and come to the conclusion that this discovery will spell the end of the power of the church and the privileged classes, and he decides to act. He meets at night in the church (I recognized with surprise that these pictures were taken in the glorious double cathedral of Erfurt in Thuringia) with a young prince of the blood whom he had succeeded in having appointed as assistant to the pseudo-Kammerer. (This is obviously a typical occupation for a German prince.) Here in the dark sacristy the plot is hatched. The prince (or the priest?) proposes to pseudo-Kammerer that he announce his glorious discovery at a formal University meeting, and the scientist gladly accepts. During the following night the priest and the prince enter pseudo-Kammerer's laboratory to which the prince has the key, as he poses as the scientist's devoted collaborator. They open the jar in which the proof specimen of salamander is kept in alcohol, and inject the specimen with ink. Then follows the scene at the University meeting. All the professors and the president appear in academic robes; the young scientist is introduced and makes a brilliant speech announcing the final proof of the inheritance of acquired characters. When the applause has ended, the priest (or was it the assistant? I am quoting entirely from memory) steps up, opens the jar, takes out the salamander and dips it into a jar with water. All the color runs out of the specimen. An immense uproar starts and pseudo-Kammerer is ingloriously kicked out of the University as an impostor. Sometime later we see the poor young scholar walking the streets and begging with an experimental monkey who had followed him into misery. Nobody thinks of him any more, when one of his former Russian students arrives and tries to call on him. Finally she succeeds in finding him completely down and out in a miserable attic. At once she takes the train to Moscow and obtains an interview with Lunacharsky (this is the scene where he appears in person) who gives orders to save the victim of bourgeois persecution. Meanwhile, pseudo-Kammerer has sunk so low that he decides to make an end of it. The very moment he tries to commit suicide the Russian student returns with Lunacharsky's message and prevents him from taking his life. The last scene shows a train in which pseudo-Kammerer and the Russian savior are riding east and a large streamer reads "To the land of liberty."

The editor suggests that there is a possibility that Kammerer's suicide gave genetics in Russia a reprieve, that no Russian geneticist could be found to play the role designed for Kammerer, and that it was not until Trofim Lysenko, who had achieved sudden prominence in 1928, could be developed into a willing and effective tool that genetics was finally destroyed.

DARWINISM

Late in 1859 Charles Darwin published *The Origin of Species by Means of Natural Selection: or, the Preservation of Favored Races in the Struggle for Life.* Here he established the theory of evolution on a scientific basis and, as a result, evolution was often called "Darwinism," a usage now archaic in biological circles. Darwinism has also been used to designate Darwin's explanation of evolution and, in this sense, is synonymous with natural selection. This second usage is also disappearing, primarily because our knowledge of the workings of natural selection has increased so greatly during the last fifty years that it is more accurate now not to label it with Darwin's name but to reserve the term for natural selection as it was understood at the time of Darwin. In Russia, however, Darwinism includes both the above meanings, and in addition all other beliefs and hypotheses which Darwin accepted. It is thus a complex of many doctrines. It is self-consistent, for Darwin was a reasonable man; and it is harmonious with the scientific knowledge of the middle nineteenth century, for Darwin kept himself well informed. When Darwin accepted Lamarck's inheritance of acquired characters, Lamarckism became a part of Darwinism, according to this Russian usage.

Natural selection, so intimately connected with the name of Charles Darwin, can be described very briefly by listing its postulates as follows: all animals and plants differ from one another, no two are exactly alike; more individuals are produced than can exist in the available space, and consequently there is a keen competition, a struggle for existence; the weakest or the least adapted are eliminated and the fit survive; the survivors form the basis for new variations which, selected long enough, produce new species.

A primitive natural selection was accepted in classical times by Empedocles, Epicurus, and Lucretius, who used it to explain the existence of adaptation through the elimination by nature of the unadapted. Certain other aspects of natural selection were recognized during the seventeenth and eighteenth centuries, and in the nineteenth it was used to explain evolution by Wells (1813), Matthews (1831), Wallace (1858), and Darwin (1858). Both Wallace and Darwin were led to an understanding of natural selection by reading Malthus. Both were aware of the changes wrought in domestic animals and plants through artificial selection, and Malthus' description of population pressure gave them a clue as to how nature selected. Malthus had emphasized the fact that population tends to increase faster than the food supply, that more individuals were produced than could reach maturity. The production of individuals in quantities too great for the available sustenance is practically universal in nature except for that portion of the human race which practices birth control. Population pressure (including the pressure of predators) is thus a major selecting agent in all species of animals and plants except in plants growing in semi-arid and arid regions, where the limiting factor is an unfavorable environment, and in animals living under extremely unfavorable ecological conditions. Population pressure is most severe, perhaps, upon the plants in a tropical rain forest. Natural selection, however, is not always dependent upon population pressure. It also operates in the absence of population pressure through the elimination of unfavorable variations and lethal and semilethal mutations. Thus it checks any increase of unadaptation.

Natural selection has also been applied rather naïvely to competition *within* human society, an application called Social Darwinism. This subject has recently been treated in detail by Hofstadter.[9]

According to Ashby,[10] Social Darwinism in Russia is "inadmissible," which means that the Authorities say to Darwinism, "Thus far shalt thou go and no farther." In medieval theology, when such an arbitrary limit was set to human thought, the subject, exempted from reason, was apt to be labeled a mystery and revered as such. Today in Russia a simple *verboten* is enough. Darwinism in Russia is also limited by the fact that a Greater Than Darwin pointed out Darwin's "error" in accepting Malthus' population pressure as a selecting agent. This Authority was none other than Friedrich Engels himself. To quote from Lysenko (page 101): "In his time Darwin was unable to free himself from the theoretical mistakes which he committed. These errors were discovered and pointed out by Marxist classicists."

In conclusion, it should be noted that Karl Marx endorsed Darwinism and that in Russia any scientist would be "mistaken" if he rejected any aspect of Darwinism which has not been rejected already by those Great Men who outrank Darwin in the

[9] Richard Hofstadter, *Social Darwinism in American Thought, 1860-1915*. Philadelphia, 1944. The rather absurd misunderstanding of natural selection by the Social Darwinians is beautifully exposed by the author. He also shows, perhaps unintentionally, that Social Darwinism was finally rejected without its biological aspects being understood. He really demonstrates that natural selection could be accepted as a basis for organized society by sociologists, or it could be rejected by them as not applying to human beings. A scientific evaluation of this aspect of natural selection was apparently just too complex for the adherents and opponents of Social Darwinism.

[10] Eric Ashby, *Scientist in Russia*. New York, 1947.

Communist hierarchy, such men as Marx, Engels, Lenin, and Stalin.

MENDELISM

Mendelian genetics, accepted by the biologists of the world since the beginning of the twentieth century, was officially rejected by Russia in August 1948. In the Russian press it has been consistently attacked and misrepresented. In fact, the Mendelism described in the following *Pravda* articles bears so little resemblance to the science taught in the universities of the world (including those in Russia until 1948, see page 143 ff.) that it is necessary to describe it here so that the half-truths and complete falsehoods can be recognized when they are encountered later. Mendelism is labeled formalistic, impracticable, sterile, idealistic, reactionary, anti-democratic, and by any other adjective which is now used in a derogatory sense in the land of the Soviets. Needless to say, our description of Mendelism will serve only to give the reader a glimpse of the subject. For obvious reasons, only those aspects of the subject which suffer the more frequent misrepresentations can be mentioned here.

In 1865 Mendel published his paper on plant hybridization. Few botanists read it at the time, and those few did not realize its significance. As biological knowledge accumulated, however, and as plant hybridizers searched the literature, Mendel's classic was found. In 1901 three botanists independently announced its discovery, and the science of genetics received a stimulus which insured its rapid growth.

Mendel had crossed a number of different types of pea, i.e., a tall plant with a short plant, a green pea with a yellow pea, a smooth pea with a wrinkled pea, etc. He recorded that the hybrids uniformly resembled one parent to the exclusion of the other, i.e., the hybrid plant was tall, not short; all the hybrid peas were yellow, not green; not wrinkled, smooth; etc. Such observations had been made intermittently during the preceding hundred and fifty years.[11] He next inbred the hybrids and discovered that both parental types appeared in the following generation, plants tall and short, peas yellow and green, etc., an observation that had also been recorded several times. Mendel took a third step, this one of the utmost importance. He counted the different types of contrasting characters and found that they occurred in the definite ratio of three to one. He studied a total of 19,959 plants, distributed through seven different experiments,

[11] H. F. Roberts, *Plant Hybridization before Mendel*. Princeton, 1929. Conway Zirkle, *The Beginnings of Plant Hybridization*. Philadelphia, 1935.

each of which dealt with a single pair of characters. The character reappearing most frequently (the dominant) occurred in between 73.7% and 75.9% of the progeny, while the character reappearing least frequently (the recessive) occurred in between 26.29% and 24.1%. Combining the experiments, the dominants constituted 74.9% of the whole, the recessives 25.1%. This regularity in the hybrid progeny gave the clue which finally led to the understanding of the basic hereditary mechanism.

Other simple ratios, in different types of crosses, were soon discovered such as 1 to 1 and 1 to 2 to 1. When two pairs of characters were involved, the simplest ratio of the four types produced was 9-3-3-1, with three pairs it was 27-9-9-9-3-3-3-1, with four pairs it was 81-27-27-27-27-9-9-9-9-9-9-3-3-3-3-1. Cases were found where dominance was incomplete and the hybrid was intermediate between the parents. Here in the second hybrid generation three types appeared, the two parental types and the intermediate. When many factors were involved, the numerous types could be fitted into a simple distribution curve. By 1910 such progress had been made that Nilsson-Ehle in Denmark, working with wheat, and E. M. East in the United States, working with Indian corn, proved that what had hitherto seemed to be a blending inheritance was really Mendelian. Many Mendelian factors were involved and, although the ratios were complex, they could be handled by very simple mathematical tools.[12] The theoretical and practical importance of this step was great, for it brought all heredity, except the relatively unimportant cytoplasmic, into the Mendelian picture.

It is necessary at this point to call attention to a persistent misrepresentation of Mendelism by the Russians. Aside from such epithets as "idealistic" and "metaphysical" attached to any aspect of Mendelism that they do not like, they frequently try to equate all Mendelism with the 3 to 1 ratio. Lysenko in particular is a consistent offender, and the reader will have no trouble in recognizing passages in his work where he satirizes this ratio. It is as if he had mastered the multiplication table as far as 2 x 2 and then discovered to his horror that there were problems to which 4 was not the answer. To be consistent with his reaction to genetics, he would then have to denounce the tables as a fraud perpetrated on the workers by their capitalistic exploiters.

The discovery of Mendelian factors, later called genes, had immediate consequences of great theoretical and practical importance. The genes, alterable by mutations, gave the evolution-

[12] This is very clearly explained in a paper by R. A. Emerson and E. M. East, "The Inheritance of Quantitative Characters in Maize," *University of Nebraska Research Bulletin*, No. 2, 1913.

ists the exact hereditary units they needed for the variations which were selected by nature. New forms well adapted and valuable need not be swamped when bred back into the general population. Evolution was finally put on a quantitative basis by the geneticists, but the evolutionists and geneticists had to learn a certain amount of mathematics in order to keep up with their growing science. Those unable to learn the mathematics simply could not understand the modern advances made in the subject. Incidentally, there is overwhelming evidence that the simplest mathematics is quite beyond Lysenko's reach.

The practical application of Mendel's discovery to agricultural problems was immediately apparent. If genes could maintain their identity in hybrids, it became relatively simple to combine useful traits which had hitherto existed only in different races and varieties. All that was needed was to cross the stocks and select the progeny which had the proper combination of genes. These combinations could be reproduced indefinitely, and new races and varieties made to order. Excellent accounts of this application of genetics to improving food plants and domestic animals are in the United States Department of Agriculture *Yearbooks* for 1936 and 1937 (a total of 2,685 pages).

The most spectacular application of genetics to increasing our food supply was the creation of hybrid corn. Indian corn turned out to be an exceptionally favorable plant for the study of Mendelian factors, and today we know more about its genes than we do about those of any other plant. Thirty-five years ago G. H. Shull, E. M. East, and a little later D. F. Jones, East's student, worked out the genetic principles involved in the increased yield of hybrid corn. By 1920 heterosis (the name coined by Shull for hybrid vigor) appeared as a heading in the textbooks of genetics. In the twenty years following, the U. S. Department of Agriculture, state experiment stations, and seed companies developed different strains suitable for different regions, and today, by planting hybrid corn, the yield in the United States is increased by nearly half a billion bushels annually.[13]

In the papers translated from *Pravda,* Mendelism is pictured as a formalistic, academic game which occupies the time of scientists who could spend their energies more profitably by working to improve the food supply of the people. It is true that organisms most suited for genetic investigation are often without economic importance. It was a fortunate accident, however, that many crop plants furnished excellent experimental material so that important advances in pure and applied science occurred to-

[13] Richard Crabb, *The Hybrid Corn Makers.* New Brunswick, 1947. I. Bernard Cohen, *Science, Servant of Man.* Boston, 1948.

gether. New and higher-yielding varieties, relatively immune to the common plant diseases, appear every year. Practically all the crops now grown in the United States consist of varieties which have been synthesized by the Mendelians.

Attention should also be called to the fact that the denunciation of genetics by the Communists is never accompanied by data of any kind.

VERNALIZATION

"Vernalization" is the English equivalent of the hard-to-pronounce "iarovization." As originally used, the term referred to treatments of germinated seeds which altered the direction of their subsequent growth. It has been extended to include also treatments which break the rest periods of seeds so that by speeding up their germination less time is needed between seeding and harvesting. Many seeds will not grow when they are first shed. Some, which are shed in the fall and germinate in the spring, have to be subjected to the cold of winter before they will sprout. By shortening the growth period it becomes possible to grow crops farther north where the limiting factor is the length of time between the last killing frost in spring and the first killing frost in autumn. The effective means used to secure this desired end is to soak the seeds in water and then chill them until the water about the seeds is frozen.[14] This is particularly effective with wheat.

There is no logical connection between vernalization and genetics. The topic is included here because Lysenko, the spearhead of the group who destroyed genetics in Russia, rose to eminence in this field and, to protect himself, has had to claim that the effects of vernalization are inherited. In Russia he is also credited with discovering the principles of vernalization and inventing the technique of accomplishing it (see page 37). This is his claim to greatness: by vernalization he "changes" winter wheat to spring wheat and accomplishes other transformations.

Lysenko's first contribution on the subject was published in 1928 (Azerbaijan Plant Breeding Station *Bulletin* #3, English Summary). Ten years earlier, however, G. Gassner (*Zeit. Bot.* 10:417-80. 1918) had both used the methods and recorded the results. Actually vernalization was old hat. It was well under-

[14] The reader is referred to the following works: H. H. McKinney, "Vernalization and the Growth-Phase Concept," *Botanical Review* 6:25-48. 1940. A. E. Murneek and R. O. Whyte, "Vernalization and Photoperiodism," *Lotsya,* Vol. I. Waltham, Mass., 1948.

stood in the United States before the Civil War and is not a twentieth-century Russian discovery. The three following quotations are offered here in evidence:

In 1837 an anonymous gentleman from Tennessee published in *The Cultivator* (Vol. 4, No. 4, p. 64) under the heading, "A Suggestion for the Coming Year," a method for sowing winter wheat in the spring:

> In early winter the seed grain is put into casks and water enough added to soak and cover it. It is then exposed so that the water becomes frozen, and it is kept in this state as far as practicable until the soil is fit for its reception in the spring. It is well known that the operation of frost upon the seed of winter grain has the same effect as if it is sown in autumn.

In 1846 Richard L. Allen published *A Brief Compend of American Agriculture,* and issued a second edition in 1847. In 1849 its title was changed to *The American Farm Book,* which went through many editions. The following quotation is from page 128 of the edition of 1850:

> The only division necessary for our present purpose is of the winter wheat *(Triticum hybernum)* and spring or summer wheat *(T. aestivum)*. The former requires the action of frost to bring it to full maturity, and is sown in autumn. Germination before exposure to frost does not, however, seem absolutely essential to its success, as fine crops have been raised from seed sown early in the spring, after having been saturated with water and frozen for some weeks. It has also been successfully raised when sown early in the season, while the frost yet occupied the ground.
>
> Spring and winter wheat may be changed from one to the other by sowing at the proper time through successive seasons, and without material injury to their character.

In 1858 J. H. Klippart published *An Essay on the Origin, Growth, Diseases, Varieties etc. of the Wheat Plant* (Ohio State Board of Agriculture, *Annual Report for 1857* 12:562-816. 1858). From page 757:

> To convert winter into spring wheat, nothing more is necessary than that the winter wheat should be allowed to germinate slightly in the fall or winter, but kept from vegetation by a low temperature or freezing, until it can be sown in the spring. This is usually done by soaking and sprouting the seed, and freezing it while in this state and keeping it frozen until the season for spring sowing has arrived. Only two things seem requisite, germination and freezing. It is probable that winter wheat sown in the fall, so late as only to germinate in the earth, without coming up, would produce a grain which would be a spring wheat, if sown in April instead of September. The experiment of converting winter wheat into spring

wheat has met with great success. It retains many of its primitive winter wheat qualities, and produces at the rate of 28 bushels per acre.

As an agricultural practice in the United States, however, vernalization was soon abandoned as obsolete. The same results were obtained better and without the trouble by breeding varieties of wheat which did not need it. These new races were obtained by the well-tested method of selection, sometimes with and sometimes without previous hybridization. Today about 360 races of wheat are grown in our country, so that there are forms suited to each ecological condition.[15] There are rumors (see page 72) that the practice of vernalization has also ceased in Russia. The official explanation is that the plastic nature of the wheat plant has been so changed through the process that vernalization is no longer needed. There are no records, however, of the most elementary scientific precautions having been taken in reaching this conclusion. Lysenko's seed were always mixed (page 76); indeed he emphatically denies the existence of pure lines, perhaps for self-protection. When a wheat mixture is grown under marginal conditions, some of the strains will be eliminated while others more suited to the conditions will thrive. Thus, the results do not necessarily prove that one variety changes into another but that one variety may replace another. To show how well this principle was understood, another short passage from Klippart will be quoted. The editor also wishes to call attention to the fact that this quotation is a hitherto overlooked description of natural selection which was published a year before Darwin's *Origin of Species.* From page 775:

> The more prolific varieties of wheat when mixed with the less prolific, bringing forth proportionally more grain, will in a few successive crops give a preponderance of the more prolific varieties and in the end supersede entirely the others, and thus without one variety being transformed into another, the character of the crop may be entirely changed in a few years, by the presence at first of so small a number of grains of the less desirable but more prolific varieties that their existence was unnoticed in the seed altogether.

Unfortunately many laymen are still impressed by Authority in science. Authority, of course, reigns supreme in Russia. It is not without its followers elsewhere, and many people are impressed with Lysenko's views in genetics because of his presumed accomplishments in physiology. These prior remarks on vernalization are quoted here because of their bearing on a particular question. Is it possible that Lysenko's originality is on a level with his logic?

[15] U. S. Dept. of Agriculture *Technical Bulletin,* No. 795, 1942.

3

WHY WAS GENETICS DESTROYED?

There is no easy answer to the question which heads this section. For us to learn just why the Russian authorities destroyed genetics we should have to look inside their minds and perhaps follow other devious trails. It is obvious that the reasons which have been published are not the real ones, but what *are* the real ones? Here we can only guess, but we have a number of facts upon which to base our speculations. Darlington has suggested a very interesting reason for the liquidation of genetics (page 72), and we can begin by presenting his statement. We must remember, however, that in interpreting the Russian actions we are not concerned primarily with what genetics actually is but with what the Great Men in the Kremlin believe it to be. To quote Darlington:

> The rise of Hitler to power gave new life to the forces working against western science [in Russia] in general and against genetics in particular. Hitler's doctrine was founded on giving a distorted predominance to a distorted genetics. His theory assumed the permanent, and unconditional, and homogeneous, genetic superiority of a particular group of people, those speaking his own language. The easy retort was obviously to repudiate genetics and put in its place a genuine Russian, proletarian, and if possible Marxist, science. For this purpose very little research was necessary: the classical personalities and achievements of Timiriazev and Michurin were there ready to hand. All that was needed was to discover a new prophet of Marxist genetics or Soviet Darwinism. The prophet was found in Trofim Dennissovitch Lysenko.

There is much to indicate that Soviet hostility to Mendelian genetics is due at least in part to the fact that genetics is concerned with problems of inheritance in man as well as in plants and animals. The Russians, like many others who are deeply involved with simple views of social problems, just do not want biological complications. We have already noted that they have exempted human beings from Darwinism as emphatically as have the American Fundamentalists. To quote further from Darlington (page 70):

> A Government which relied on the absence of inborn class and race differences in man as a basis of its political theory was naturally unhappy about a science of genetics which relies on the presence of such differences

amongst plants and animals as the basis of evolution and of crop and stock improvement.

Certain other bits of evidence fit into Darlington's views. Muller states[16] that the Genetics Congress scheduled for 1937 was canceled because some geneticists insisted on their right to give papers on human heredity.

This step [the canceling of the congress] was taken only after the Party had first toyed with the idea of allowing it to be held with the omission of all papers on evolution and on human genetics in spite of the fact that many foreign geneticists had intended in such papers to attack the Nazi racist doctrine!

We also have evidence from other sources that human genetics was a dangerous subject in Russia. The attack made on the Medico-Genetic Institute of Moscow in 1936 is one example. We find additional confirmation in the acts of the Russian censors, in that the reports of the genetic debates of December 1936 were published (in the Russian edition) with all references to man expurgated, and the book itself was soon banned. Furthermore, Newman[17] stated that work on identical twins at the Maxim Gorky Institute was suddenly terminated in 1939 and nothing further was heard of the psychologists in charge of the work. To quote Muller further:

What causes the Communist officials to push Lysenkoism so strongly? To me, the answer is obvious: it is the type of mind that sees things as only black and white, yes and no, and so cannot admit the importance of *both* heredity and environment. Believing that it has found the complete answer to all the world's ills, through its particular way of manipulating environment, the Communist Party regards as a menace any concept that does not fit patly into its scheme for mankind. The genes do not fit into that concept, in its opinion, hence the existence of the genes must be denied.[18]

These explanations of the destruction of genetics are probably not complete, nor do they exclude other and possibly more important factors. Darlington himself has noted that the Russians are not overburdened with consistency. In the land of the Soviets, human beings had been excluded from Darwinism through the simple fact that Engels had decreed precisely how far Darwinism could extend. Mendelism could be limited just as easily and, with

[16] H. J. Muller, "The Crushing of Genetics in the USSR," *Bulletin of the Atomic Scientists* 12 (4): 369-71. 1948. Morton Stark, "The Stockholm Congress," *Journal of Heredity* 39:219-21. 1948.

[17] H. H. Newman, *Multiple Human Births.* Chicago, 1940.

[18] H. J. Muller, "Back to Barbarism Scientifically," *Saturday Review of Literature,* Dec. 11, 1948.

Stalin alive and available, the means of excluding mankind from such unwelcome hereditary limitations was so easy that it should not have required a second thought. Then, too, not all geneticists agree that the genic differences between human classes are important. If the Soviet authorities hated such genetic hypotheses, there was no need to destroy genetics in general; as has been pointed out by Dobzhansky (personal communication), they would only have to withdraw their support of the geneticists who held such views, and give generous support to their opponents. We cannot doubt that this procedure would have worked.

Michael Polanyi[19] offers an alternative explanation. He states that when the final authorities who control science are not scientists themselves, sooner or later quacks will flourish and ultimately will dominate the field. Even with the very best intentions, those who are not scientists cannot decide scientific questions and cannot, in the last analysis, even choose between scientists and charlatans. It is only when science is free and autonomous, when enlightened scientific opinion is the court of last appeal, that impostors can be exposed as they arise. The plausible oversimplifications of quackery have always had a great popular appeal, otherwise quackery would not flourish. Concerning the genetics controversy, Polanyi states:

> The demonstration given here of the corruption of a branch of science caused by placing its pursuit under the direction of the State, is, I think, complete. The more so—I wish to repeat—as there is no doubt at all of the unwavering desire of the Soviet Government to advance the progress of science. It has spent large sums on laboratories, on equipment, and on personnel. Yet these subsidies, we have seen, benefited science only so long as they flowed into channels controlled by independent scientific opinion, whereas as soon as their allocation was accompanied by attempts at establishing governmental direction they exercised a violently destructive influence.

There are undoubtedly a number of other contributory factors in the Russian reaction against genetics. Genetics denies the inheritance of acquired characters, and this type of inheritance promises so much so easily that it has always been a favorite of those who want to make over mankind in a hurry. Consequently, for a very long time it has been very popular with certain types of reformers and revolutionaries. Condorcet (1794) based his belief in the "perfectability" of man upon it. William Godwin (1797), the noted liberal, endorsed the belief, and Lamarck (1809) himself held that the inheritance of acquired charac-

[19] Michael Polanyi, "The Autonomy of Science," *Mem. & Proc. Manchester Lit. & Phil. Soc.* 85:19-38. 1943. Reprinted in *Scientific Monthly* 60:141-50. 1945.

ters provided a means which, in time, could eliminate human inequalities. The popularity of the belief in Russia as early as 1926 (see page 18) was shown by the Kammerer tragedy. Any group of scientists who denied such inheritance was bound to be unpopular with those who held an escapist belief in a quick reconstruction of man and his works. Such folk could easily miss the implications of acquired characters accumulating through past ages. In the Party line press of the United States the inheritance of acquired characters is now endorsed in articles that should be fascinating for any scientist who would like to learn how "science" is presented to the faithful.[20]

One corollary to the inheritance of acquired characters has consequences not at all to the liking of the Communists. It has been pointed out by Muller *(op. cit.)* that apparently this implication of the doctrine had been overlooked:

> In the official [Communist] view, individuals or populations which have lived under unfavorable conditions and have therefore been physically or mentally stunted in their development, would tend, through the inheritance of acquired characteristics, to pass on to successive generations an ever poorer hereditary endowment; on the other hand, those living under favorable conditions would produce progressively better germ cells and so become innately superior.

Thus the inheritance of acquired characters provides a *modus operandi* for establishing superior and inferior races and classes. The *cause* of such differences, as here described, is almost antipodal to the cause assigned by the Nazis, but the end result is the same. The fact that this inequality is denied in other portions of Communist dogma does not destroy the logic of the corollary. The Party was undoubtedly ignorant of the fact that anthropologists first used the inheritance of acquired characters, operating over many generations, to explain *racial* differences (Voss 1658, Rumpf 1721, Mitchell 1744, Maupertuis 1745, Camper 1764, Josephi 1770, Hunter 1775, Herder 1784, Sömmering 1785, and Smith 1787). It was *later* used to explain species differences (E. Darwin 1794, Lamarck 1802, Chambers 1844, Spencer 1864, C. Darwin 1868).[21] It is clear that the Central Committee never understood the evolutionary implications of the inheritance of acquired characters or how incompatible it is with an equalitarian philosophy.

Not all of the plausible causes for the destruction of genetics have yet been listed. The basic cause which the editor would as-

20 *The Daily Worker*, Oct. 17 and Dec. 26, 1948. *Soviet Russia Today*, Jan. 1949.

21 Conway Zirkle, "The Early History of the Inheritance of Acquired Characters and of Pangenesis," *Trans. Amer. Phil. Soc.* 35:91-151. 1946.

sign will probably seem trivial to most people, if not actually preposterous. Soviet science, as Darlington has pointed out, seems to be derived from a canon which included Marx, Engels, and Lenin and those they endorsed. Lenin had blessed a nurseryman, an importer of plants named I. V. Michurin (1855-1935), who thus achieved a high rank in the Communist Apostolic Succession. Michurin was a firm Lamarckian and, in his dotage, when his Bolshevik record and Lenin's approbation had raised him beyond the criticism of mortals, he became violently anti-Mendelian (Dobzhansky, personal communication). As a consequence, Michurinism is never criticized in Russia and Michurin's doctrines are regularly greeted with lavish praise, as will be seen in the translations from *Pravda*. Even Vavilov dared not contradict any of his pronouncements and, as the lesser geneticists also wanted to live, any honest discussion of the basic problems of genetics was out of the question in the scientific academies of Russia. The style of the geneticists was thus definitely cramped, and their speeches and arguments sounded depressed and futile. Consequently, in the "socialist competition" for position and power in the biological field the type of winner could be predicted easily. No honest scientist could ever be as good a Michurinist as Lysenko, Prezent, and their followers.

The editor is aware that the question why genetics was destroyed in Russia has not been answered. It will probbaly puzzle historians of science for many years to come. Perhaps there is some simple reason which we have overlooked. It may be that the destruction of genetics is only part of a larger tragedy, involving the general degradation of science. Psychology has practically disappeared in Russia since the war, and in the fall of 1948 physics was under fire. Statisticians dare not publish conclusions unpalatable to the Central Committee. It is possible that authoritarianism simply cannot allow the existence of the intellectual standards of free scientific inquiry.

II

The Decline of Genetics, 1939-40

1

EVENTS OF 1939

The International Genetics Congress, whose proposed 1937 meeting in Moscow had been canceled by the Russians, met in Edinburgh during the week of August 23-30, 1939. A more dramatic moment or a less auspicious time for a scientific meeting could hardly have been chosen. The Second World War was just beginning. The German geneticists who had been allowed to attend were recalled before the sessions were over and left hurriedly in a body. Many delegates from other European countries also rushed home, but the Americans, whose return passage had been booked in advance, stayed until the Congress ended. A few sailed on the *Athenia,* which was torpedoed in mid-Atlantic, the first liner to be sunk in the war.

N. I. Vavilov had been elected President of the Congress and had accepted the honor, but later he had to decline as neither he nor any other Russian geneticist was allowed to attend. A prescience of war was probably not a cause of the Russians' failure to meet with their colleagues, for the war which was beginning was only the Imperialistic War, a conflict in which Russia was not involved. It would not become the Great Patriotic War until June 1941, nearly two years later.

Less than six weeks after the International Congress ended, the Russians held a congress of their own. It met from October 7 to 14 in Moscow and was sponsored by the editors of *Under the Banner of Marxism.* Fifty-three members participated in the discussions, and the greater part of their talks was published in the sponsoring magazine. Three of these addresses, but slightly abridged, have been republished in English by *Science and Society: A Marxian Quarterly* (4:183-233. 1940). The three participants whose speeches were chosen for translation were N. I. Vavilov to represent the geneticists, T. D. Lysenko to give the viewpoint of the Michurinists, and I. M. Polyakov, who took an intermediate position. This impartial presentation of the speeches of the scientist, the charlatan, and the straddler constitute valuable source material for anyone who wishes to correlate what the speakers said with their ultimate fate.

It is generally admitted that Vavilov's address was one of the poorest he ever made. His own arrest was only ten months in the future, and it is certain that the cold war on him was taking its toll, that he saw how things were going and knew that he was doomed. Lysenko, on the other hand, was on his way up and was destined to replace Vavilov and to become the great biological authority in Russia. Polyakov was to discover, in the fullness of time, that there was no safe middle course and that safety can be secured in Russia only by being on the winning side. He recanted but still lost his job in 1948 (page 289).

The comments on the Russian conference by the editors of *Science and Society* are well worth reading in the light of subsequent events. They expressed the hope that their translations would make the commentators in the United States less inclined to inject suppositions of political motives into Russian scientific controversy. They also called attention to a long critique and evaluation of the conference written for *Under the Banner of Marxism* by Professor M. B. Mitin, head of the Philosophical Institute of the Academy of Sciences, a critique which expressed more than any other the attitude of the Soviet Government. In addition the editors stated that Mitin gave a "most balanced judgment of the contending parties," the "substance" of which was published in *Pravda* on December 7, 1939. (A translation of this article constitutes section 2 of this chapter.)

The editor wishes to record his complete disagreement with *Science and Society* concerning Mitin's balanced judgment. In his opinion the only judicious aspect of Mitin's *Pravda* contribution was its tone. Early in the article Mitin mildly rebuked both Vavilov and Lysenko on some very minor points, in a manner that sounded most impartial. He thus established the tone of his article and apparently succeeded in deceiving those who could not recognize the technical evidence of his bias. From this point on, he endorsed every claim of Lysenko and praised him at every opportunity. On the other hand, he assailed Vavilov's position throughout both by innuendo and by actual falsehoods. In fact, the editor intends to use Mitin's article as an example of an official, systematic, and skillful misrepresentation of genetics and geneticists.

This task is not too difficult, because Mitin was handicapped by the fact that he knew genetics too well to be able to achieve the vague confusion of Lysenko, some of whose meaningless pronouncements are beyond refutation. Mitin did very well, though, considering the disadvantages of his intelligence and knowledge, since he was skillful enough to confuse just those complex aspects of biology which are somewhat difficult for the untrained to grasp, even without any gratuitous misinformation. His chief weakness in supporting the Party line in science seems to be his habit of thinking clearly. Actually he made his misrepresentations precise enough to be refuted in not too many words. Outside of Russia, of course, he can do no harm, for geneticists can recognize his tricks as they read them. In Russia, however, questions of genetics are not settled by geneticists but by the Party, and Mitin has risen high in the Party where, now that genetics is dead, his opportunity for injuring other sciences is great.

2

TOWARD THE ADVANCEMENT OF SOVIET GENETICS[1]

M. B. MITIN

(Abridged Translation)

[*The editorial footnotes appended to this article are exceptionally numerous. The reason is that it is economical both of time and space to concentrate a large proportion of the necessary footnotes to the translations from* Pravda *in this, the first one. The basic types of misinformation and unscientific reasoning appear over and over again in these articles, and it would be very tedious to expose them every time they occurred. Once they are pointed out, however, they are easy to recognize and hence it will not be necessary to repeat the editorial comments in the subsequent articles.*]

Not long ago a meeting called under the auspices of the journal *Pod Znamenem Marksizma* was held for the purpose of dis-

[1] Printed originally in *Pravda,* Dec. 7, 1939. This translation appeared in the *American Review on the Soviet Union* (2:37-48. 1940), published by the American Russian Institute, Inc. We are indebted to the Institute for permission to republish.

cussing some basic problems of genetics and selection which have been the subject of protracted discussion among advocates of different tendencies in these sciences. Fifty-three speakers were heard at the meeting, among them, people of rank and importance: scientists, academicians, practicing geneticists, teachers.

The aim of the present article is to acquaint the reader of *Pravda* with certain results of this meeting at which some of the most important questions of biological science were discussed.

Academician Lysenko, as a consequence of unflagging work carried out over a period of many years, is producing practical results which possess great economic significance. He discovered and put into practice a method of vernalizing grain.[2] He worked out a method of combatting the degeneration of the potato in the south, thanks to which we have found it possible to provide our south with good potatoes. Lysenko is developing a fruitful, Michurinian method of vegetable hybridization. It is evident that in all these results, there is still much that must be checked, made more exact, developed further.[3] But it is equally evident that in the results of Lysenko we possess much that is new, valuable and useful.

The achievements of Lysenko also possess great theoretical significance. They carry science forward, and work out new implications of a number of fundamental doctrines of Darwinism. Let us take, for example, his theory of stages in the growth of plants. In what does the scientific significance of this theory consist? Lysenko proved the existence of a series of consecutive stages in the growth of plants. He showed that such characteristics of yearly growth as summer or winter maturation, adaptability to a short or long day are not changeless or unmodifiable. Such a view as his suggests that it might be possible to transform a plant from a winter to a summer variety and vice versa. Lysenko has shown that it is impossible to understand the characteristics of a plant species apart from the external conditions under which the species exists, that each plant requires specific and definite conditions in the course of its development. The working out of these important doctrines is without doubt a considerable contribution, both to applied agriculture and to biological science.

The basic theses of Academician Lysenko that the plant, both in respect to species and individual forms, is not something unchangeable; that it has no absolutely unchanging characteristics;

[2] This claim of Lysenko's discovery of vernalization is false (see page 26).

[3] This very mild criticism of Lysenko is to balance an equally mild criticism of Vavilov (see note 4).

that all its characteristics are subject to change in the course of the evolution of the organism; that in this evolution not only the sex cell or its component elements play a part, but the organism as a whole and the cell as a whole; that in the course of its evolution, not only the external characteristics of the organism but also its hereditary base are subject to change; that man can, by means of his active intervention, his deliberate creation of definite environmental conditions, influence the evolution of the organism, directing its development into channels useful to him—all these important theses of Academician Lysenko are not only in harmony with the teachings of Darwin, but represent a further development of a whole series of Darwinian doctrines as well as an extension of the method of dialectical materialism in biological science.

Lysenko does not look upon science as upon some precious rarity which exists only to be admired, and performs no further function. No, science is something necessary for life, for practice, for the modification of nature.

And how do matters stand with the opponents of Lysenko, with the representatives of so-called formal genetics?

Academician N. I. Vavilov is rightly revered as a scientific authority. He has contributed considerable work which possesses a great deal of significance for biological science—work involving the creation of a seed collection which is the only one of its kind in the world. What Academician Vavilov has created has not only practical significance, but also general theoretical interest. But precisely because Academician Vavilov is a leading scientist, our collective society has the right to make demands upon him that are by no means small: namely, to come closer to life, to practice, to bridge the gulf between science and practice.[4]

Academician Vavilov spoke, in his report of world science, of world genetics. He gave an interesting review of what is taking place in that field at the present time. However, two things stand out in his review: first, a certain uncritical acceptance of many world authorities in the field of genetics, an absence of a critical attitude towards them in cases where such would have been fully warranted. In the second place, Academician Vavilov put forward the thesis that genetics has been developing along Darwinian lines and is now Darwinian. This thesis represents an incorrect view of the situation, for it shows an undiscriminating attitude towards the actual process of growth of the science under

[4] Mild as this criticism is, it is very useful in giving an impartial tone to Mitin's article but it has no basis in fact. Vavilov spent the greater part of his energy and time in the practical application of genetics to agriculture, very little in research in pure genetics.

discussion.[5] Is there perhaps one continuous line of development within genetics? Are there in this science no mutually opposing influences, no struggle between antagonistic elements?

Actually, there is taking place in genetics a sharp struggle between Darwinism and anti-Darwinism. There is the case of Punnett, certainly one of the leading geneticists in "world science," who is anti-Darwinist. In 1938 he announced in the article "Forty Years of Evolutionary Theory" that "the beginning of our century is connected with the inauguration of Mendelism, and the appearance of the work of Bateson and De Vries can be counted as a landmark, as registering the end of the Darwinian era and the beginning of the Mendelian era in evolutionary theory."[6]

Shull, the well-known American geneticist, moves in the direction of a denial of the creative role of selection,[7] criticizes the Darwinian theory of mimicry.

Heribert Nilsson,[8] prominent Swedish geneticist and selectionist, has gone as far as a denial of the theory of evolution and selection, and openly calls for a return to Linnaeus and Cuvier. He holds that "Darwinian evolution has shown itself unable to meet the requirements of life."

Finally, take Morgan—central authority of contemporary genetics. Here is what he writes in his book, *Experimental Foundations of Evolution:* "Since all this appears debatable,

[5] Note the use of the term "incorrect." Neither evidence nor data of any kind are presented. Mitin is obviously a higher authority than Vavilov so he can label any statement of Vavilov "correct" or "incorrect."

[6] There is, of course, no struggle over organic evolution among the geneticists. They all accept it. The advantage Mitin gets by substituting Darwinism for evolution is easily seen. But even so he makes no attempt to show how Mendelian *theory* is incompatible with Darwinian *theory*. He merely relies on quotations taken out of their context. His dishonesty can be exposed most easily by quoting the opening paragraph of Punnett's article. It is on page 191 of *Background to Modern Science* (Cambridge, 1938): "In the minds of almost all who are familiar with his name Charles Darwin stands for the concept of Evolution, for the essential unity pervading the diversity of organic form. Hence in assigning a time limit to the 'Darwinian era' we may clearly take as our starting-point 1859, the year which saw the publication of *The Origin of Species*. So long as men believe in evolution, so long in that sense will the Darwinian era continue to be with us. But since for our present purpose some circumspection of its boundaries is necessary I will take as the latter limit the year 1900, the year in which the rediscovery of Mendel's work began to lead to that re-orientation of the biological sciences which is still going actively forward."

[7] The joker here is the word "creative." Darwin himself did not look upon selection as creative. He supposed that nature merely "selected" the variations which were already present. Mimicry is a very minor problem in evolution. Its explanation is now an open question, and it makes very little difference whether Darwin's explanation is accepted as he gave it or whether it will have to be modified.

[8] **A systematic botanist.**

should we still make use of the term 'natural selection' as a part of the mutation theory, or should we exclude it because now it no longer possesses that meaning which the followers of Darwin gave to his theory?"[9]

All this testifies to the fact that it is impossible to speak of a "world science" as of something single, whole, and continuous. The development of science in class, in bourgeois society proceeds through contradictions. In science we find a struggle between advanced and backward tendencies. Consequently it is necessary to approach the course of developments in "world science" with a discriminatory attitude.

Holding as we do a high opinion of the role and significance of the works of Academician Lysenko, conscious of the positive values in the practical and theoretical results of his movement, we must not, however, refrain from criticism of the specific exaggerations and faults of the movement nor from criticism of individuals connected with it.

One of the central questions discussed at the meeting was the question of Mendelism. Darwin set up a scientific biology. On the basis of enormous factual material he founded the theory of evolution in the organic realm. But that does not mean that there were no shortcomings in Darwin's work or that he elucidated all the problems of organic evolution. On the contrary, precisely on that level to which Darwin elevated himself, there stands a whole series of new problems which Marx and Engels even then very keenly observed, forecasting the direction of the further development of Darwinism.

Darwin's theory of evolution is constructed on the basis of the recognition of variation, inheritance, and selection as the controlling factors in the evolution of organic species. Out of all these factors of evolution Darwin placed the chief emphasis on *selection* and worked out, in the main, the problem of the actual operation of selection. The problem of the causes of individual variations of organisms Darwin did not work out.

After Darwin, science amassed a huge amount of new factual material. The structure of Darwinism expanded to colossal proportions and the study of inheritance and variation developed in vigorous fashion. Contemporary genetics, together with a whole series of related sciences, addressed itself to these questions.

Since the beginning of the twentieth century a series of outstanding discoveries has been made in research on the mechanism

[9] A quotation taken out of context. Morgan merely raises the question whether "natural selection" should be considered a part of the "mutation theory" or separate from it.

of inheritance and the phenomenon of variation. The foundation of the science of the cell—cytology—was laid. The study in all its details of the structure of the cell, of the role and significance of the chromosome apparatus in general, particularly in connection with the transmission of hereditary characteristics—such investigations certainly belong in the class of outstanding and undeniable scientific achievements.

From this point of view let us examine particularly the question of the Mendelian laws. There is no doubt about the fact that Mendel discovered certain laws in connection with the inheritance of definite characteristics: the phenomenon of segregation among hybrid offspring, the well-known mathematical law of this segregation, the relative independence of certain hereditary characteristics. Mendel's discoveries in the field of inheritance were then connected with processes within the cells of the organism, in particular in the sex cells. [Mitin here discusses the attitude of the biologists Timiriazev and Michurin towards Mendel's laws, citing passages from their works. Associating himself with their attitude, he speaks of "a just, clear, scientifically objective evaluation of Mendel and Mendelism, avoiding as one-sided either an uncritical adulation of Mendelism or a wholesale rejection of its significance in the science of inheritance." He then proceeds.]

If even a very small part of nature's processes is brought to light by any scientist whomsoever, why should we not profit by such knowledge? We must. The discoveries of Mendel unquestionably reveal certain aspects of the phenomenon of inheritance, and thus embody knowledge of which we must avail ourselves.[10]

Let us pass over to the following question which also was subjected to long and serious discussion. That is the question of the significance and role of cytogenetical investigations in connection with the study of problems of inheritance.

It is evident to all that in recent times a whole series of important studies have been made in connection with the cell. The study of the role of the cell and of its separate elements in connection with the development of different characteristics of the organism is flourishing. For instance, the significance of the differentiation of specific cell structures in regard to the isolation of species in the process of evolution is being clarified. Cytogenetic studies permit us partly to lift the veil on the question of the inheritance of properties of organisms. Much is cleared up in

[10] This paragraph shows that in 1940 the Soviet authorities were not ready to discard Mendelian genetics. But in 1948 (see page 279) Mendelian genetics is presented as a reactionary view which had never been acceptable to the Party.

relation to the characteristics of sex. These are all matters possessing great scientific significance. It is necessary to carry on elaborate research work in these fields.

But at the same time that important achievements are being won in the field of cell research, there have grown up in this science certain empty, metaphysical conceptions which are obstacles to its further development. A very serious example of such barren conceptions which hinder the further development of genetics is the metaphysical theory of the gene[11] which at the present time occupies something of a central place in formal genetics, and which, as it figures in contemporary genetics, is basically incompatible with the theory of evolution.

By "genes" in contemporary genetics is understood some sort of factor (or complex of factors) placed in the chromosomes of the nucleus of the cell and determining the characteristics of the adult form. The chromosome itself is represented as a complex of such particles. This is what geneticists have in mind when they speak of "the material basis of inheritance," when they speak of contemporary genetics explaining materialistically the phenomenon of inheritance.

However, we must be much more cautious than to take such a definition of gene as "materialism." We must remember that there is a mechanistic, vulgar, *metaphysical* materialism irreconcilable with the theory of evolution and there is dialectical materialism, founded on the theory of evolution.[12]

Geneticists, rejecting the theory of preformation which asserts that in the seed or embryo of an organism are contained in complete but smaller form all the characteristics of its adulthood, emphasize that in the embryo, in its chromosomes, there are not the characteristics of the adult individual, *but there are only genes, factors, the possibilities or potentialities of characteristics which become actualities only in case there are present certain definite external conditions.* In the embryo of the black rabbit, they say, *there is no "black color," but there is a factor which can lead to "black color"* if the embryo evolves under suitable conditions.[13]

[11] An example of "metaphysical" as a derogatory adjective. The statement is meaningless.

[12] This paragraph is quoted on page 9 to show how the meaning of words is destroyed.

[13] Here Mitin begins an extremely clever bit of misrepresentation. He does not say that geneticists believe in preformation, but notice his implication in what follows. He tries to make the concept of the gene appear to be a modern revival of the doctrine of preformationism which was disproved two hundred years ago. He then proceeds to refute the abandoned doctrine.

Notwithstanding such a reservation, the net result of this view is that the chromosome is a combination of factors, already containing the possibility of certain specific adult characteristics. This means that in one way or another the formal genetics theory involves the view that there exists a complete correlation between the characteristics of the adult individual and the factors, capacities, contained in the embryo.[14] This is what contradicts the principle of evolution. Particularly it contradicts the law of Darwin and Haeckel, according to which the individual in its embryological development reproduces in shortened form the path of development of the species to which it belongs.[15]

The mistake of the doctrine of the gene consists in this, that it is too simple, it crudely and mechanically connects the characteristics of the *adult individual* with capacities in the *embryo* of this individual.[16] Formal geneticists do not give due consideration to the fact that between the embryonic cell of the individual and its adult stage lies a *long path of evolution*.[17] In the evolution of the species this path sometimes takes up tens and hundreds of thousands of years. In the evolution of the embryo it is reduced to minimum time. But it is important to note that in this evolution—from the embryo to the fully formed organism—there takes place various *qualitative transformations*. For the evolution of the species to be reproduced in that of the embryo means that in this latter evolution the embryo undergoes changes of its characteristics, its structure and of all the material elements contained therein. Now when the embryo reaches the stage where it has already acquired traits of its adulthood, it possesses characteristics for which, earlier, there were not and could not be any special factors or "genes." Here we may witness how the doctrine of the gene forces us into a whole series of absurdities. If the human embryo in its intra-uterine development reaches a stage where it is like an amphibian, possessing gills, then becomes like a lower mammal, then like its simian ancestor, we must conclude, according to the gene theory, that in the chromosomes of its parents there were factors determining all characteristics of the

[14] The misleading words here are "factors, capacities." The word should be "potentialities." The "capacities" of the embryo are presented as preformed characteristics of the adult individual. If the genes were presented honestly as units, probably giant protein molecules, whose interaction determined the limits within which the adults' characters must develop, then the issue would not be obscured.

[15] This so-called biogenetic law is not as simple and universal as it was thought to be by Haeckel.

[16] This simplified view is of Mitin's own creation. It is not held by geneticists.

[17] The word here should be "development," not "evolution."

organism at all stages of its embryological development and at all stages of its development as an adult.[18]

The doctrine of evolution and the phylogenetic law testify convincingly against the theory that in the embryo there are *already* factors determining the characteristics of the adult. It is obviously an over-simplified conception.[19]

The theory of evolution says that in the course of evolution a series of *qualitative transformations* are possible and that accordingly there can arise anew in the adult individual and in the embryo as well, at the different stages of its development, characteristics for which in the embryonic cell there were originally no genes, no corresponding factors.[20] Living organic matter possesses an attribute not yet explained by science: to reproduce in the development of the embryo the qualitative transformation gone through by the species.

Let us take as an example the evolution of the butterfly. In the beginning it is a fertilized egg cell, then a caterpillar, then a chrysalis, then a colored butterfly. By virtue of what does this lump of embryonic matter from which the butterfly emerges accomplish such transformations? It would be an extremely simple and every mechanistic explanation to say that the *embryo* of this living creature possesses genes of the caterpillar, the chrysalis and the butterfly.

It would be much more complex and much more correct to approach this problem from the point of view of the theory of evolution. It is a problem we can solve only by means of a study of the embryo in its development, taking into account the evolution of the individual and the species as a whole.

Here is an instance of the complexity of the theory of evolution: it allows for *qualitative changes* in the development of the embryo. But the theory of genes permits only the *extension,* the further *manifestation* of that which was originally placed in the embryo.[21]

It is sometimes said that nowadays the adherents of the theory of genes acknowledge their relative modifiability. If such is the case, why do they not explicitly criticize the claim, so pervasively asserted in so-called "classic" genetics literature, that the gene is unalterable?[22]

[18] Only by considering genes as preformed adult characters could this paragraph make sense. Actually particular genes have been recognized through their effects on many different stages of development.

[19] See note 16.

[20] See note 18.

[21] See note 18.

[22] No claim has ever been made by a reputable geneticist that the gene is unalterable. In fact, particular genes at loci where no mutations have occurred cannot be studied genetically.

Here is an example. In the article "Genetics" printed in 1929 in Volume XV of the *Bolshaia Soviet Encyclopedia,* Academician Vavilov writes: "The gene represents a definite, invariable unit of heredity which can be compared with the atom in chemistry and physics. . . . The genes are transmitted from generation to generation without changing their nature."[23]

If this older statement of the matter in respect to the unalterability[24] of the genes is now rejected by Academician Vavilov, then it is necessary openly to criticize it and thus clear the deck for further work. We ought to give up the theory of genes, metaphysical from start to finish, and free ourselves from something that stands in the way of an evolutionary treatment of the cell. It is necessary, on the basis of the wealth of factual material which has accumulated, to set up a whole series of new problems which will raise our Soviet genetic science to a new level, corresponding in fact to that height on which our country stands in comparison with the capitalist world.

Let us take the following group of questions which once again bear witness to the blind alley in which formal genetics finds itself and to the importance of avoiding these blind alleys.

Take for instance the concepts of genotype and phenotype. Unquestionably, they possess genuine scientific significance. It is necessary to distinguish between the hereditary element of the organism and its outward development. Dialectics teaches us to distinguish, for example, substance from its manifestations. We utilize, and cannot fail to utilize, the concepts of genotype and phenotype in our practical and theoretical work. Formal geneticists, however, think of these two concepts as belonging to two different realms, separated by a metaphysical Chinese wall.[25] The *theoretical* result of all this is simply *metaphysics*. The *practical* result is fruitless, hampering theory which we must thoroughly root out.

The divorcement in principle set up in contemporary genetics between genotype and phenotype, between so-called mutations and modifications, between internal and external factors of development does not exist in nature.[26]

[23] Enough of this quotation has been left out to enable Mitin to present a half-truth. Genes are remarkably stable units and are not altered by such regular experiences as duplicating themselves everytime a cell divides or by passing from one generation into another. If they were not so stable, no form would ever breed true. However, genes mutate often enough to provide the raw material for evolution, and enough mutant forms accumulate in our domestic animals and plants to enable us to improve them by artificial selection.

[24] See note 22.

[25, 26] The editor believes that Mitin is here indulging in a little "deadpan" humor. What would the readers of *Pravda*, or the Central Committee for that matter, know about phenotypes or genotypes? Why should he not see what he could do with two such

Formal geneticists have fallen into the habit of accusing of Lamarckianism[27] all those who in the slightest degree acknowledge the action of the surrounding life conditions of the organism on its hereditary elements. They accuse of Lamarckianism even Darwin himself, who acknowledged the influence of climate and food, exercise and non-exercise and in general environment in the broadest sense. Formal genetics, not being able to find a bridge between organism and environment, between mutation and modification, between genotype and phenotype, is cut off from the road leading to the actual influencing of the nature of organisms through manipulating the external conditions of their existence.

It is time to put an end to such metaphysics. It is necessary to address ourselves with the utmost attention to those practical achievements with which actual life confronts us, in particular to those which are connected with the works of Academician Lysenko. It is time to implement enthusiastic declarations of attachment to dialectical materialism by a genuine attempt to set these problems before ourselves in our scientific fields.

The meeting showed that on the part of many representatives of so-called formal genetics there is an egregiously disdainful attitude towards new phenomena, new sides of life—an attitude unsuited to Soviet scientists. It is necessary to struggle against a certain pontifical attitude, professional cliquishness, aloofness, hostility to the new, distaste for self-criticism, which may be found among formal geneticists.

But at the same time it is also necessary to struggle against even the slightest manifestation of an improper attitude towards our Soviet intelligentsia, contributing to the welfare of socialism.[28]

We can and indeed must have disagreements in science. We can and must have our theoretical quarrels. We should therefore rebuke and exclude from science any administrators who would hinder such developments.

---

beautiful words? Here was an opportunity for any connoisseur of the confusing. It was too bad that the misrepresentation is so easy to refute. The terms are very easy to explain. All individuals in a genetic experiment who look alike are called a phenotype. But all that look alike do not necessarily breed alike or have the same genetic formula. All individuals who have the same genetic formula constitute a genotype. Thus, far from being separated by a Chinese wall, phenotype and genotype are intimately connected. Each phenotype is composed of one or more genotypes. The two words are just useful terms, generally used in teaching beginning students elementary genetics.

[27] This paragraph is evidence that in Russia in 1939 "Lamarckianism" was a bad word. See page 156 for contrast.

[28] This paragraph is also evidence that in 1939 the Soviets were not ready to discard genetics. See also note 10.

3

THE ARREST OF N. I. VAVILOV, 1940

No happening in the Russian genetics imbroglio has aroused more interest and concern than the arrest and death of Vavilov. His arrest was reported erroneously as early as 1936 by the *New York Times* as the result of a mistake in the transmission of a Moscow dispatch. As late as 1947 it was still not admitted by Professor J. B. S. Haldane.[29] Between these dates numerous statements can be found both affirming and denying it. Geneticists outside of Russia had not been able to get in touch with Vavilov for some years, and his name has been dropped from the list of recent members, both living and dead, of the Russian Academy of Sciences. Since the war, however, more details have seeped through the Iron Curtain. The reader is referred to the article on Vavilov written by Professor Th. Dobzhansky (reprinted here on pages 80-89). Subsequently, Professor Dobzhansky has received more information which he has graciously placed at the disposal of the editor. The following account is made possible by this information.

In the summer of 1940 Vavilov organized an expedition to Galicia and Bukowina, then occupied by Russia. The object was to collect samples of indigenous agricultural plants before they had been displaced by the new varieties imported from Russia. The expedition was organized in Kiev but was delayed and sabotaged by local officials who knew that Vavilov had fallen from grace. Finally the expedition started and reached Lvov. The personnel was divided into several groups which went into various parts of the territory to be studied. News soon reached the collectors that Vavilov had hurriedly left Chernowiz, the capital of Bukowina, for Moscow. The news was astonishing, since Vavilov was supervising the work closely and his absence consequently was unexpected. In a week or so the collecting groups came together and learned via the grapevine that Vavilov had been arrested and was not expected back "soon." The discouraged expedition was disbanded. There seems to have been no local speculation as to the cause of Vavilov's arrest. In fact, at this time, people in Russia did not talk about such matters and

[29] J. B. S. Haldane, *Science Advances*. New York, 1947. See also review of *Science Advances* by I. Bernard Cohen in *Isis* 38:255-56. 1948.

tended to avoid even mentioning the names of people who had "disappeared."

The date of Vavilov's arrest is August 20, 1940, plus or minus a week. His subsequent fate and death are described on pages 80-89.

III

A Controversy with a Twisted Ending 1943-47

1

THE PARTICIPANTS AND THEIR CONTRIBUTIONS

The controversy presented in this chapter has aspects which are probably unique in the history of science. Certainly none of those who took part in it could have foreseen how far it would extend, how long it would last, or what the outcome would be. The beginning was quiet enough.

On November 7, 1943, the National Council on American-Soviet Friendship held a congress to celebrate the tenth anniversary of American-Soviet Friendship. A number of papers were presented at the Science Panel, which was a part of the celebration. Among the papers was one entitled "Soviet Biology" by Professor L. C. Dunn. It was the sort of polite, friendly paper suitable for the occasion, but it started a chain reaction. Although presented in 1943, it described conditions in Russia as the author saw them during his visit in 1927. At this earlier date the cold war on genetics had not begun. Research in genetics was getting under way, and the geneticists were busy and seemingly well pleased. This period in the history of Russian genetics should not be forgotten, and we should not let the later debacle obscure the fact that once genetics lived in Russia. Professor Dunn's address was published in *Science* on January 28, 1944, and later the same year in *Science and Society* (8:64-66. 1944).

In Russia after 1937, however, genetics suffered a number of blows; Professor Karl Sax called attention to them in a later issue of *Science* (April 14, 1944), and listed a number of omissions in Professor Dunn's paper. The controversy was on. Professor Anton R. Zhebrak of the Timiryazev Agricultural Academy of Moscow joined in the discussion. He endorsed what Dunn had written, and stated that Sax "gives an incorrect estimate of the general position of biology in the USSR" (*Science* 102:357-58. October 5, 1945). Sax replied (*Science* 102: 648-49. Dec. 21, 1945), and the argument apparently came to an end.

For two years the matter rested. Then it seems that either the Party line shifted in genetics or some member of the Lysenko clique discovered Zhebrak's paper. At any rate there was suddenly an astonishing display of fireworks with Professor I. Laptev in the role of chief pyrotechnist. His paper in *Pravda* on September 2, 1947, ended the controversy with a well-planned, climactic explosion. It will be the concluding section of this chapter.

Before presenting Laptev's display, however, attention should be called to three papers which appeared during the two quiescent years. These contributions were not directly involved in the controversy, but they were on the debated subject and were quoted copiously by Laptev. The first of these in point of time was "Soviet Science and Dialectical Materialism" by John Somerville (*Philosophy of Science* 12:23-29. 1945). The second paper, "A Revolution in Soviet Science" by Professor C. D. Darlington (*Discovery* 8:40-43. Feb. 1947; reprinted in the *Journal of Heredity* 38:143-48. 1947), is very pertinent to the argument. The last of the three, "Work of Soviet Biologists" by Dr. N. P. Dubinin (*Science* 105:109-12. 1947), gives an excellent factual account of the accomplishments of Soviet geneticists. Dubinin made a major mistake, however, in not burning any incense on Lysenko's altar. In fact, he did not even mention Lysenko's name. This was a very costly blunder, as will appear later.

Perhaps no controversy has ever been brought to a conclusion as decisively as this one. When Laptev had finished there was nothing that anyone else could say. The incident was closed. But the incident itself may have been a trigger which set off a greater reaction.

2

SCIENCE IN THE USSR: SOVIET BIOLOGY[1]

DR. L. C. DUNN

At the time of the tenth anniversary of the October Revolution in 1927, I was in Moscow; I awakened each morning in the little glass-sided cupola on top of the palatial and elegant mansion which had now become the Institute of Experimental Biology. My first impression was one of familiarity, of at-homeness, for this was a genetics laboratory, filled with the sights and smells associated with the little fly, *Drosophila,* which breeds in its thousands in the milk bottles of fermenting food which line all the genetics laboratories in the world. But in the farther distance, through the windows, were the spires of Moscow, and these and the physical world they represented were utterly strange and new to me.

This alternation of strangeness and familiarity must have struck many American visitors to Russia, and it persists when we try to examine the scientific achievements of the Soviet Union or indeed of any country not our own. For any modern science is in some sense the same wherever we find it, a part of one interconnected whole resting on common basic principles, with a common past and a common future, and it is artificial and deceptive to try to break it into separate national entities. And yet, just as the history of science consists in part of the achievements of individuals, so also it rests on the contributions of groups of persons with common purposes and common methods, and oftentimes the character of these groups is determined by the physical, economic and social milieu. It was unquestionable that the society behind Soviet biology was very different from that found in Europe and America, and this, together with the temperament, traditions and outlook of the Soviet scientists lent a distinctly Russian flavor to their joint work. There was too a kind of revolutionary tinge about their manner of approach to some of the problems of biology. Whereas Westerners were inclined to go in through the traditional front door, our Soviet colleagues seemed at times to break in through the back door or even to come up through the floor.

[1] Published originally in *Science* 99:65-67. 1944. We are indebted to *Science* for permission to republish.

Thus it comes about that it is possible to speak of "Soviet" biological research and to single out for comment a few of its characteristics. I should like it understood that I do this from a very limted knowledge of Soviet biology which covers a vast field and that I can speak with confidence only about work which is closely related to my own field.

The qualities in Soviet biological research which have struck me most are first, from the purely scientific side, its vitality and activity, and the atmosphere of eagerness, modernity and novelty which has surrounded it. To the outsider looking in it has had aspects of youth and originality which have never attached, for example, to the scientific renaissance which was taking place at about the same time in Japan. In the second place, one Westerner at least has noted the peculiar and almost paradoxical combination of philosophical and theoretical impetus with which practical purposes are pursued. On the organizational side, the peculiarity of Soviet biology is of course that it is centrally planned and administered, chiefly through the Academy of Sciences, that its purpose is not only to discover new knowledge but to penetrate the whole life of the community. It is thus of very great scope both with regard to the numbers of persons engaged in it and in its institutional and geographic connections.

The great vitality of Soviet biology is nowhere better evidenced than in my own field of genetics and its close relative, cytology. Here there is no doubt that the most important contributions have been coming from the U. S. A. and USSR, and in the number of workers, of institutes and in quality of work these two countries are comparable. Genetics has been recognized in Russia as one of the disciplines underlying agriculture and medicine and has received a large measure of support. Professor Koltzoff, director of the Institute of Experimental Biology in Moscow, told me of how he traveled to Leningrad during the famine of 1920 with Lenin and some other members of the Central Executive. Lenin was to urge upon the responsible committee the diversion of some of the funds set aside for famine relief to the construction of a research institute for seed selection and plant breeding. "The famine to prevent," said Lenin, "is the next one and the time to begin is now." He carried his point and there was built with emergency funds the great Institute of Applied Botany which under the direction of Nikolai Vavilov became the center of the great plant breeding and seed selection service in the world. Vavilov himself became the world authority on the history of crop plants.

In 1921 also the American geneticist, H. J. Muller, took to Moscow strains of the vinegar fly *Drosophila* and there grew up the greatest center of theoretical research in this field outside of the United States. Although the impetus came from America, the Soviet workers soon took their own line, and there was founded under Tschetverikoff the important new field of population genetics for the study of the distribution of new hereditary characters in nature. In the hands of Dubinin, Timofeef-Ressovsky and Dobzhansky, the latter now in the United States, this developed into the most important new experimental approach to the problems of evolution. Out of Soviet genetics have come also new ideas of chromosome structure, of the origin of mutations and new ideas on the arrangement and relations of the hereditary particles, the genes, by very many workers. By 1940 Moscow had in fact become one of the most important centers of work of this kind.

The comparative scope of genetical work in the USSR and the esteem in which it is held is illustrated by the fact that in this third year of Russia's participation in the war, she is still the largest foreign subscriber to the chief American scientific journal in this field. More copies go to the USSR than to all other foreign countries. Moreover, a standard American textbook which appears in the United States in editions of 2,000 copies is printed in the USSR in editions of 15,000.

The spirit in which the Soviet scientists carried on their studies in the difficult days just after the Revolution is again in evidence today. After the fall of Kiev I received a letter from Professor Gershenson, director of the Genetics Institute of the Ukrainian Academy of Sciences at Kiev, telling of the destruction of the institutes and the loss of the libraries. The personnel had been evacuated to two small towns, one in the Urals and one in Turkestan, and there they were continuing their work. They needed, he wrote, recent American publications and some stocks of *Drosophila*. We are now collecting books and journals to send to replace those destroyed by the Nazis.

There are today literally hundreds of trained genetical investigators in the USSR, certainly more than in any country outside of the U. S. A. They had already outstripped the Germans in this field even before the advent of Hitler put the quietus on German genetics. Soviet theoretical genetics has developed in close connection with practice, especially with agriculture and medicine, and has been continually aware of the relationship between its own theoretical structure and the social theory on which the development of the USSR has itself been based.

Other aspects of biological research which have shown great expansion and activity include the remarkable outburst of exploring and collecting zeal by which the animals and plants of the vast and varied autonomous republics and of China became known. This began immediately after the Revolution and has resulted in a great enrichment of the museums and in works of first-class importance by both zoologists and botanists. At the same time there began the development of institutes of experimental biology from which has issued important work in experimental morphology, on the analysis of growth, in endocrinology, in physiology and in biochemistry. In the latter field for example the discovery that the contractile protein, myosin, which is the basic component of muscle, acts as an enzyme, was of first-rate significance; while we all have increasing reason now to remember that much of the pioneer work on blood transfusion and plasma storage was done in the Soviet Union. The methods of artificial insemination, now used extensively in Europe and America, were developed almost wholly by Russian workers. It was estimated in 1941 that 50,000,000 farm animals in the Soviet Union alone had been produced by artificial insemination.

In recent years there have appeared in the Russian scientific literature new hybrid names indicating the fusion of independent scientific disciplines to focus on problems which transcend particular fields. Such is biogeochemistry, as conceived by Vernadsky and his group of Biogeochemical Institute of the Academy of Sciences at Leningrad. Vernadsky took as his field the distribution of chemical elements due to living organisms in the biosphere and has added greatly to our knowledge of the chemistry of alluvial soils and the chemical composition of organisms.

Biological progress in the Soviet Union has not been achieved without cost and sacrifice. At a time when food was scarce they still spared some for their experimental animals and for costly scientific equipment. They took the means to build up science literally out of their necessities, not, as we have done, out of our surplus; and they had only themselves to look to, for the great foundations, which poured their funds for pure research so generously into Germany and western Europe, were never able to make the same arrangements in or for the Soviet Union. Yet in the USSR was what in 1927 seemed to me to have been the greatest potential source of new scientific strength in the old world.

Some part of the cost was paid too in the creation of the central control of science which led to what we call red tape and the Russians "spoiling paper," and to an appearance of arbitrariness whenever decisions are made by a central authority. I have no doubt there was wailing and gnashing of teeth on the part of the individual investigators when before the war one of the great biological institutes was suddenly moved from near Leningrad to Moscow, but in view of what the Germans did to Leningrad, I cannot believe that the regret of those biologists has survived to the present. These costs, together with other and greater ones, have been and are being paid, and we can now see that not only Soviet citizens but those of all countries stand to reap the benefits.

The progress of biological research in the Soviet Union has taught us a very valuable lesson. It is that control and organization of science by and for the whole community does not kill the scientific spirit or initiative nor submerge the individual scientist in a dead level of anonymity. Great individuals have arisen in Soviet biology, fine discoveries have been made and continue to be made even in the midst of war. Ivan Pavlov, one of the greatest of Russian biologists, began his scientific life under the old régime, but he lived to refute both in word and in deed the dire prophecies of those who said that great scientists and a vital and vigorous science could not survive in a socialist state.

For the sake of biological science itself, we biologists should use all our efforts to see that the barriers which separated Soviet biology and biologists from us should never again be allowed to prevent the free flow of persons and ideas, both scientific and social, on which the progress of science and of society depends.

3

SOVIET BIOLOGY[2]

KARL SAX

The eulogy of Soviet biology published in a recent issue of *Science* did not present a realistic survey of the present situation. The author did not discuss the most significant trend of biological research in the USSR—the subservience of science to social and political philosophy. It is important that this aspect of Soviet science should be generally known and understood because it is not confined to Russia. It could happen here.

The work of Russian geneticists, plant breeders and cytologists, during the early years of the Soviet régime deserves the highest praise, as does the Soviet government for providing such generous support for scientific work. About ten years ago the influence of Soviet political philosophy began to appear in bio-

[2] Published originally in *Science* 99:298-99. 1944.

logical science, culminating in a public controversy regarding the relative roles of environment and heredity in 1939. Much of this controversy has been published in this country,[3] and a more damning indictment of the new Russian biology would be difficult to imagine. Vavilov, while recognizing the effect of environment on development, emphasized the progress of genetics and the role of heredity in plant breeding. Lysenko, on the other hand, upheld the Lamarckian (in his words "Darwinian") concept of variation, and rejected Mendelian heredity and genetics as a science. He also claimed that "any hereditary properties can be transmitted from one breed to another without the immediate transmission of the chromosomes." His discussion of "vegetative hybrids" resulting from grafting might well have been written in 1800; his views are neither original nor heterodox, but merely archaic.

Lysenko's attitude towards genetics presumably was influenced by his earlier work on vernalization. A winter wheat, which differs from a spring wheat by a single genetic factor, can be grown as a spring wheat if the seed is moistened and chilled for several weeks before planting. This discovery was made in the United States before the Civil War. Vernalization is also said to hasten the maturity of other crops. This technique has been tried in many other countries without sufficient success to warrant commercial utilization, but it has been used extensively in Russia.

Lysenko and his associates seem to have convinced the political authorities that only environmental effects are of value in plant improvement. Since 1939 the Soviet plant-breeding journals have been filled with articles by Lysenko's disciples, but we hear nothing from Vavilov, Karpechenko, Navaschin and the many other able scientists who are responsible for building the foundations for Russia's plant-breeding program. A few examples of the recent plant-breeding methods are typical of the new order. In one case scions of a yellow-fruited tomato variety when grafted on a red-fruited stock are said to produce progeny segregating for fruit color. Dolgusin claims that halves of the same plant, when grown under different environmental conditions and then crossed, produce progeny of increased vigor and fertility. The ovules are supposed to select the pollen grains most favorably affected by the environment. This selective power of the gametes is referred to by Lysenko as "marriage for love."

There are several reasons for the suppression of genetics in the USSR. A nationalistic attitude is reflected in Polyakov's reference to genetics as a "foreign science." Another factor may have

[3] *Science and Society,* A Marxian Quarterly. Summer, 1940.

been a reaction to the distortion of genetic principles by the Nazis in their myth of racial superiority. The primary factor, however, appears to have been based upon political philosophy. It is particularly significant that the Lysenko-Vavilov controversy was reviewed by Mitin, head of the Philosophical Institute of the Academy of Science, and that he "more than other commentators" expressed "the attitude of the Soviet government."

Our admiration for the Russian people and the military might of the Soviet Republic should not blind us to the fact that science has not been free in the totalitarian states where science must conform to political philosophy.

4

SOVIET BIOLOGY[4]

ANTON R. ZHEBRAK

Science has published recently two articles on Soviet biology. The article by Professor L. C. Dunn (99:65-67. 1944) gave, in the main, a correct impression of the position held by Soviet biology in world science and of the innovations in the development of science brought about by the new social structure of the USSR. The general tone of Professor Dunn's article shows his profound understanding of the real nature of the new Russia and his desire to further friendly relations between the scientists and the governments of our two countries.

Professor Sax in his reply to Professor Dunn 99:298-99. 1944) gives an incorrect estimate of the general position of biology in the USSR. He raises a number of political questions and maintains that Soviet political philosophy suppresses science in the USSR and that biology is not free to develop. His statements are based entirely on what he believes to be the present position of his own branch of science, genetics, in the USSR.

He and other American geneticists may be pleased to know that his beliefs reflect a misunderstanding of the true facts, and that actually the science of genetics is making progress in the Soviet Union. A number of important genetics laboratories are doing good work: the laboratory of the former Institute of Experimental Biology, now the Academy of Sciences Institute of

[4] Published originally in *Science* 102:357-58. 1945.

Cytology (directed by Professor Dubinin); the laboratory of plant genetics at the Timiriazev Agricultural Academy (directed by the present writer); the laboratory of genetics at Moscow University (directed by Professor Serebrovsky); the laboratory of fish genetics (directed by Professor Romashev); the laboratory of genetics at the Gorky University (directed by Professor Chetverikov); the laboratory of plant selection at the Timiriazev Agricultural Academy (directed by Academician Lisitsin); the laboratory of genetics of the Ukrainian Academy (directed by Professor Gershenson); the laboratory of genetics at Voronezh University (directed by Professor Petrov), and a number of others.

At the Genetics conference held in Moscow University (December 12-19, 1944) the achievements and future tasks of Soviet and world genetics were discussed by a large gathering of scientists. A few of the papers read were those of Professor Dubinin on "Basic Problems in Genetics," Academician Serebrovsky on "Modern Trends in Genetics" and the present writer on "Problems of Polyploidy."

Important results have been achieved in the field of polyploidy, and many new plant forms have been developed. This research is little known abroad, although some of my work on the development of new varieties of wheat has been mentioned in *Science* and *Nature*. Sakharov and Lutkov have developed new varieties of buckwheat; Navashin—kok-sagyz (the Russian dandelion, as it is called in America); Lutkov—flax; Rybin—hemp, etc. The work of these scholars offers splendid prospects for the future selection of the crops concerned. Those familiar with the earlier literature of Russian genetics will recognize most of the names listed above as those of scientists who were making outstanding contributions long before the beginning of the controversy about which Sax speaks so strongly. Their present position is in itself evidence that the careers of many Soviet geneticists have not been adversely affected by the above-mentioned controversy. There have also been some noteworthy achievements in the investigation of the delicate structures and the changeability of chromosomes, the intracellular factors in heredity. Professor M. Navashin, whom Dr. Sax specifically mentions among those from whom he hears nothing scientifically, has published four articles since 1939, and in 1944 handed in two more for publication. These papers represent a continuation of the excellent work in cytogenetics for which he has for some time been highly esteemed in both the United States and the USSR. They in no way indicate that he has been forced to curtail or modify adversely his scientific work because of political pressure.

Unfortunately Professor Sax seems to be ignorant of this work, since he regards present-day Soviet biology and particularly genetics as synonymous with the name of Academician Lysenko's mistaken views and appears to be of the opinion that he is not only criticizing Academician Lysenko but also the very foundation of Soviet biology, and the Soviet Government's attitude toward science.

This grave misunderstanding may have arisen partly through the impaired communications between the United States and the Soviet Union in recent years, so that he has not been able to see the papers resulting from the work mentioned above. The actual situation is as follows.

As an agronomist, Academician Lysenko has put forward a number of practical suggestions which have been of great value to the Soviet Government. Many Soviet geneticists are, however, sharply critical of his theories about genetics, and in no way support his attempt to re-examine and discard a number of the fundamental postulates of our science.

The Soviet Government has never interfered in the discussions of genetic questions which have now been raging for some ten years. Academician Lysenko was rewarded for his work in the field of practical scientific farming and not for his views or experiments on genetics. Furthermore a number of our geneticists and plant breeders—some of whom have developed new varieties of the chief grain crops (Konstantinov, Lisitsin, Shekhurdin, Yuriev, the present writer and several others) and who have sharply criticized Academician Lysenko's views on genetics and selection—have also been decorated by the Soviet Government.

These facts should serve to show that Academician Lysenko's criticism of genetics, based as it is on naïve and purely speculative conclusions, despite the vigor of its assault is incapable of impeding the onward march of genetics in the USSR.

The fact that Academician Lysenko is director of the Institute of Genetics of the Academy of Sciences does not mean that other schools of Soviet geneticists are in any way hampered in their work. It would be wrong to deny that Academician Lysenko has influenced the development of genetics in the USSR, but this influence has been exerted in open debate between proponents of different scientific views and principles and not by political pressure, as described by Professor Sax.

The way in which science has developed in the USSR combines centralized planning with the creative endeavor of individual, decentralized laboratories. The present war has witnessed a tremendous development of science in our country, especially pure science. In the course of just over a year (up to December, 1944) four new academies were established—the Academy of Medical Sciences, the Academy of Pedagogical Sciences, the Uzbek Academy of Sciences and the Armenian Academy of Sciences. Science is also well developed in our universities and other schools of learning.

From this outline, it can be seen that science can be free in a centralized socialist state, which Dr. Sax wrongly calls totalitarian. Professor Sax does not understand the essence of the Soviet conception of the bonds between pure science, its application and philosophy, according to which the materialist philosophy of nature can only develop on the basis of the development of the various sciences; he therefore incorrectly states that in our country "science must conform to political philosophy." Because of this lack of understanding he failed to realize that the statements of Lysenko concerning the supposed refutation of Mendel's laws on the basis of dialectic materialism have little in common with the serious development of philosophy in the Soviet Union. Dialectic materialism is based on real facts and never denies them. Therefore the philosophy of dialectic materialism, when truly understood, cannot possibly hinder the development of genetics. This philosophy, on the contrary, is a powerful weapon in the hands of the scientist who has thoroughly mastered it, and one which helps him to solve the most complicated theoretical problems.

The quotation of Professor Sax from *Science and Society* reflects a mistaken view on the part of the editors of this journal. In his speech made in 1939, Academician Mitin expressed entirely his own views, and not in any way the viewpoint of the Soviet Government.

The undoubted strength and vitality of the Soviet Government is due to that of its founder, Lenin, who himself had a long training in science and the philosophy of dialectic materialism, and of its present leader, Stalin, who has continued and strengthened the scientific and philosophical basis of our State. The well-known achievements of the Soviet Union in all branches of knowledge are due entirely to the attention which the Soviet Government has paid to the development of all forms of scholarship. The progress of science and culture was a deciding factor in the victories of our armies over fascist Germany and her satellites, who had the industrial might of all Europe behind them.

Russian scientists have a high opinion of the scientists of America and the great contribution they have made to world science. They respect the democratic principles of American society which Hitlerite reaction menaced in the same way as it menaced our Soviet State principles—a higher form of democracy in so far as all State bodies are elected by the whole people in accordance with our democratic constitution. Together with American scientists we who are working in this field in Russia are building up a common, world-wide biology. We hope that this unfortunate misunderstanding of the basic ideas of our country, and of the path of development taken by Soviet science will be speedily dispelled, and that in the future the scientists of the two countries will progress together in an atmosphere of mutual understanding and comradeship.

5

SOVIET BIOLOGY[5]

KARL SAX

In his recent report on Soviet Biology Dr. Zhebrak assures us that "the careers of *many*[6] Soviet geneticists have not been adversely affected by the above-mentioned [Vavilov-Lysenko] controversy." If, as Zhebrak claims, Lysenko's "influence has been exerted in open debate between proponents of different scientific views and principles and not by political pressure" why should the career of any Soviet geneticist be so "adversely affected"? Of the three geneticists specifically mentioned in my original article Dr. Zhebrak accounts for only one. What has happened to Karpechenko, the geneticist who laid the foundation for work on allopolyploid hybrids which Zhebrak has developed so successfully? Where is Vavilov, one of Russia's greatest scientists and one of the world's greatest geneticists? Vavilov was elected president of the International Genetics Congress which met in Edinburgh in 1939, but Vavilov did not attend, and we have not heard from him since. We now have information from our National Academy of Sciences that Vavilov is dead. How did he die and why?

The American geneticists have long recognized the valuable work done in the Soviet Union and have enjoyed the most cordial personal relationships in the past, but even before the war it was difficult to maintain personal contacts. No Soviet scientists at-

[5] Published originally in *Science* 102:648-49. 1945.

[6] Italics mine.

tended the International Botanical Congress in Amsterdam in 1935 or the International Genetics Congress in Edinburgh in 1939. Perhaps lack of funds kept them at home, but China and India were represented. Isolationism in science, or in any other field, has no place in a modern world. We hope that we may soon resume communication and personal association with our Russian friends and colleagues.

6

ANTIPATRIOTIC ACTS UNDER THE GUISE OF "SCIENTIFIC" CRITICISM[7]

I. LAPTEV

Certain English and American journals with pretentious scientific names have been carrying in recent years a wrathful, slanderous, and completely unscientific propaganda against progressive Soviet biological science. The February issue of the English journal *Discovery* contains an article, "Revolution in Soviet Science" by a scientific reactionary, C. D. Darlington. This member of the Royal Society has seen fit, by a single stroke of the pen, to liquidate Russian biological science, just as the fascist politicians have in their time "abolished," on paper, the Soviet Union, transforming it into a . . . "geographic concept." Darlington has maliciously declared that Timiriazev's discoveries "are nothing but the requirements of William of Occam," of a medieval scholastic. Darlington calls Michurin "an importer of plants" from Canada and U. S. A.! He attacks T. D. Lysenko in the same style.

The slanderous article of Darlington has provoked a justified resentment among English scientists. Angry letters with sharp protests against this article, against the despicable slanderous attacks on Soviet scientists, poured[8] into the editorial office of *Discovery*. H. G. Creighton wrote about the article of Darlington: "the political irritation and the vindictive anticommunist and antisoviet wrath have blinded his objectivity and have led

[7] Printed originally in *Pravda*, Sept. 2, 1947. This English translation was published in the *Journal of Heredity* 39:18-21. 1948. We are indebted to the *Journal of Heredity* for permission to republish.

[8] A total of six letters was published in *Discovery*, five anti-Darlington, one pro-Darlington. H. G. Creighton stated that he considered genetics "reactionary."

him to some very unscientific arguments against Lysenko." Another American scientist, M. Hamilton, declared to the editor that, "a disagreement in the field of genetics cannot serve as a justification for dirty political insinuations," and that Darlington has pleased by his articles only those who "in the depth of their soul hate science and all that it has given and gives to mankind."

One can add nothing to these truthful and objective evaluations of Darlington's article. They clearly show the inner motivation of the attacks on Lysenko and other Soviet biological scientists.

Darlington is seconded in U. S. A. by a certain Professor Karl Sax, who is angered by a mere mention of dialectic materialism and its application in the field of genetics. He is displeased especially because Soviet scientists consider fideism, i.e., the "doctrine" which replaces knowledge by faith, to be foolishness. He endorses the racists demagogic assertions well known to the Soviet people, that our country is "totalitarian" and that biological science in it is "suppressed."

Karl Sax's effort appears in the American journal *Science.* It is notable that in U. S. A., as well as in England, a series of scientists have protested against Sax's reactionary political insinuations. Dunn, a professor of Columbia University, in his reply[9] to K. Sax in the same journal declared: "The progress of the biological research in the Soviet Union gives us a very valuable lesson. Great individuals have arisen in Soviet biology, valuable discoveries have been made and continue to be made now in the midst of war" (*Science,* 1944, No. 2561, p. 67).

J. Somerville, an American scientist who attended the discussion of the problems of genetics in USSR in 1939, wrote: "So long as representatives of the state act according to the principles of dialectic materialism, there can be no question of scientific conclusions made under dictation, or of an artificial limitation of scientific activity. The dialectic materialism is a philosophic doctrine which not only does not hold science in contempt but considers it the basic source of its own principles" (*Philosophy of Science,* 1945, Vol. 12, No. 1).

In the light of all this, a feeling of deep indignation is provoked by the debut of A. Zhebrak, a member of the White-Russian Academy of Sciences, on the pages of the journal *Science* with an anti-patriotic article, "Soviet Biology," directed against progressive Soviet biological science.

One could expect that the Soviet scientist, A. Zhebrak, would stand against the dirty slander on Soviet science made by reac-

[9] This statement is false. Dunn's paper was published ten weeks *before* Sax's.

tionary English and American "scientists," such attacks on Soviet science as, according to the justified remark of the English biologist Haldane, came in the past "from Hitler's friends." Unfortunately, we see the exact opposite. Under the guise of polemics with Professor Sax, A. Zhebrak has completely adopted the latter's stand. Together with the most reactionary foreign scientists, he humiliates and defames our progressive Soviet biological science and its eminent modern representative, Academician T. D. Lysenko.

First of all, A. Zhebrak has completely misrepresented Soviet biological science before the American reader. In dealing with problems of plant physiology, genetics, and selection, he has deliberately omitted to mention such leaders of science as Timiriazev, Michurin, and Williams, whose work is the basis of that of the modern Soviet investigators in this field. Not a word is said about such world-renowned scientific institutions as Lenin's All-Soviet Academy of Agricultural Sciences, the Institute of Genetics of the Academy of Sciences of USSR, and others.

Solidarity with the reactionary K. Sax is undignified for a Soviet scientist, A. Zhebrak. Writes A. Zhebrak: "Unfortunately, Professor Sax . . . regards modern biology, and particularly genetics, as synonymous with the name of the academician Lysenko. In criticizing the mistaken views of Lysenko, he apparently holds the opinion that he criticizes not only the academician Lysenko but also the very basis of Soviet biology and the relation of the Soviet government to science."

One must have lost the feeling of patriotism and of scientific honesty to declare that a scientist known to the whole world, a pioneer in the field of genetics, academician Lysenko, has no relation to Soviet biology. . . . What a feeling of patriotic indignation will be the answer to this of every Soviet peasant, every writer in socialist agriculture, who for a number of years has been getting good harvests on the basis of scientific methods developed by T. D. Lysenko!

A. Zhebrak has needed this in order that, having given over the Soviet scientist T. D. Lysenko for the amusement of the corrupt capitalist press, he might represent himself a devoted ally of Darlingtons, Saxes and their like, who chose to "criticize" T. D. Lysenko in order to fight for reactionary idealism against dialectic materialism.

The political meaning of this fight is obvious to everyone. Reactionary biologists know it well. None other than Robert Simpson, in an article directed against Lysenko and entitled "Science—According to the Totalitarian Model," has declared, "The dual role of Lysenko in politics and science calls for a fearless inquiry by scientific men into the dangers of science becoming subservient to the state. . . . It's time to break a few lances." Simpson is worried because the Soviet land has found methods to get an abundance of grain, milk, meat, wool, and fur. He is worried because capitalism is unable to give this abundance to the people.

Zhebrak, as a Soviet scientist, ought to have unmasked the class meaning of the struggle taking place in connection with problems of genetics. But, blinded by bourgeois prejudices, by a contemptible subservience to bourgeois science, he has adopted the view of the enemy camp. To please this camp, he has dishonored representatives of Russian science in a foreign journal which specializes in maligning Soviet scientists. For A. Zhebrak there exists such a thing as "pure science." Zhebrak writes: "together with American scientists, we, working in the same field of science in Russia, are building a general biology on a world-wide scale."

So, this is the source of the antipatriotism of this "scientist"! So, neither does progressive Soviet biological science exist in nature, nor is there a reactionary idealistic biology. So, there exists only a single biology "on a world-wide scale." This explains the alliance of Zhebrak with Darlington, Sax, and other obscurantists of the reactionary capitalistic camp.

A national exclusiveness is remote from Soviet scientists. In their investigations, they generalize everything progressive which is given by the foremost scientists of foreign lands. They develop science on the basis of comradely criticism and self-criticism. They are imbued with the spirit of life-giving Soviet patriotism, the spirit of Soviet national pride in their science, science which develops on the only solid foundation, dialectic materialism—science which serves the interests of the people. A Soviet scientist is proud to wage an unrelenting struggle with that "science" which is profoundly inimical to dialectic materialism, which serves not the people but the fooling of people, and the strengthening of the power of exploiters.

The antipatriotic act of A. Zhebrak is aggravated by a personal attack on the academician T. D. Lysenko. With a petit-bourgeois impertinence, he declares on the pages of an American journal that academician Lysenko was rewarded by the government "not for his opinions or experiments in the field of genetics" but only "for his work in the field of agricultural practices." Knowingly misrepresenting the orders of the Presidium of the Supreme Soviet of USSR regarding the bestowal of high government rewards on the academician Lysenko for eminent scientific attainments, for theoretical scientific works, which have a first-rate importance in the practice of socialist construction, A. Zhebrak is calumniating Soviet science by this dishonest declaration. More than that, he makes the absurd assertion that the work of academician Lysenko ". . . founded in reality on naïve and purely speculative conclusions, is unable to interfere with successful development of genetics in USSR." It follows that, according to Zhebrak, Lysenko acts as a brake on progressive development of genetics! A. Zhebrak has gone too far.

How disgusting is the role of Zhebrak in taking it upon himself to please the reactionaries of the whole world by defamation of his compatriot scientist in the pages of a foreign journal inimical to us! Only a man who has lost all sense of social duty could take such a step.

One could close the discussion of this antipatriotic attack by Zhebrak, but we have before us another issue of the American journal *Science,* Vol. 105, No. 2718, for 1947. It has an article by N. P. Dubinin (Institute of Experimental Biology of the Academy of Sciences of USSR), entitled "The work of Soviet biologists: theoretical genetics."

First of all, one is put on one's guard by the fact that this journal, inimically disposed toward Soviet biologists, gives a front place to Dubinin's article. But the answer comes easily. All the founders of Soviet biology are ignored in the article. Whom does Dubinin regard as Soviet biologists? It happens that here are lauded such "Soviet biologists" as Dobzhansky, Timofeef-Ressovsky—open enemies of the Soviet people.

Dobzhansky, for example, is now working in U. S. A. in the field of slander against Soviet biologists and "creates the style" in this anti-Soviet campaign. With a feeling of loathing and indignation one reads further in this article praises for a series of similar personages. The whole Michurinian trend in biology is ignored by Dubinin. Every Soviet biologist will resolutely rise against such an obviously wrong, clearly antipatriotic, representation of Soviet biology.

What do the above facts mean? They mean that a certain backward part of our Soviet intelligentsia still carries a slavish servility for bourgeois science, which is profoundly foreign to Soviet patriotism. We are proud of our country, of victorious socialism, and of our progressive Soviet science. It is necessary to pull out decisively and pitilessly the rotten roots of obsequiousness and slavishness toward bourgeois culture. Only in such a way can one accomplish successfully the great task set before Soviet science by Comrade Stalin, namely, to exceed in the briefest possible time the attainments of science in foreign lands.

To the court of public opinion with those who act as a brake on the accomplishment of this task, those who defame our progressive Soviet science by their antipatriotic acts!

IV

Genetics in Russia after Ten Years of Cold Official Warfare

1

THE RETREAT FROM SCIENCE IN SOVIET RUSSIA[1]

C. D. DARLINGTON, F.R.S.

The development of science in Russia since the revolution of 1917 has been made known to the outside world, in spite of a restricted intercourse, in a variety of ways. Scientific publications, usually in the earlier days with foreign language summaries; international scientific congresses and celebrations in Moscow, always well planned and provided; popular publications circulated abroad describing officially the achievements of Soviet science; finally, numerous articles and books by western writers, some of whom have visited more than one important centre in Russia, have described the work that was being done, and sometimes the work that would be done, by Soviet scientists.

The views of some of these western writers have been coloured, perhaps legitimately, by the proclaimed adherence of Soviet statesmen to a philosophy, the Marxist philosophy, which attributes a pre-eminent importance to science and hence seems to allow scientists a more respectable and even a more authoritative position in society than they enjoy elsewhere. Some visitors, too, like the English geneticist Bateson in 1924, have been prepared to excuse a certain rigour in the régime on grounds of the scientific purity of its principles and its aims. The importance that has been attached to the effective application of scientific discoveries for the benefit of the whole community has also been contrasted with

[1] Republished by permission of the *Nineteenth Century* 142:157-68. 1947.

their frustration in other countries. All scientists, indeed, were bound to rejoice that the conflict between science and society which had existed elsewhere seemed to have been finally resolved by the Bolshevik Revolution.

There have, however, been doubtful voices. Some have wanted to know why so many of Russia's most outstanding workers, both in the physical and in the biological fields, should have left their native land when that was still possible. Some, too, have asked whether the pressure for practical application and the requirement of philosophical orthodoxy might not amount to a heavy price for a scientist to pay for seeing his work made use of.

The answer to these questions can now be attempted. The development of one of those sciences lying at the root of present-day advance, namely, the science of genetics which concerns itself with heredity and evolution, has been fairly fully revealed since the end of the war. It has been described in English by a number of scientific translators, travellers, inquirers and reporters of various races and nations and political opinions. The story can therefore be told and interpreted on sufficient documentary authority.

When Lenin examined the situation of genetics, and the possibilities of its useful application, in the Soviet Union in 1921 he found it comparing not unfavourably with that in the rest of Europe. The confusion of parties and opinions found elsewhere was to be sure enhanced by political controversy. Timiriazev, a leading plant physiologist of radical politics, had been engaged in sustaining Darwinism, which is a necessary part of Marxist doctrine (and by Darwinism he correctly meant to imply evolution by natural selection) equally against anti-evolution and against anti-selection. With this second heresy Bateson had implicated Mendelism and the new science of genetics. In a preface of 1905 we therefore find Timiriazev denouncing Bateson as the 'head of clerical Anti-Darwinism.' Apparently in the heat of argument Mendel's theory of heredity could be connected with his abominable profession as a monk.

In spite of the emotional views he expressed in his old age, Timiriazev recognised the uses of Mendelian segregation. One of the great difficulties of Darwin's theory of natural selection, in his own opinion, had been that differences blended in inheritance. New characters would therefore be lost before they could be selected for their advantages by nature. To overcome this difficulty Darwin had latterly fallen back on the so-called Lamarckian theory. This was the ancient superstition that a changing environment could directly change heredity and so mould and adapt races and species without the help of natural selection. Mendel, however (unknown to Darwin), had shown that different types did not blend in heredity. On the contrary, they were passed on, unchanged by one another or by the environment, from generation to generation and, in consequence, mixed parents have children, brothers and sisters, different from one another. The process by which they do so is known as Mendelian segregation. It was Timiriazev who first pointed out that this segregation brings out the differences for selection to act upon and hence that Mendelism is not merely consistent with Darwinism but provides its necessary foundation. It is not surprising therefore that, after the death of Timiriazev, Lenin chose a Mendelian to take over the development and application of plant breeding, and indeed of agricultural science generally, in the Soviet Union.

It was in 1921 that Lenin appointed N. I. Vavilov,[2] then aged 36, as President of his new Lenin Academy of Agricultural Sciences and Director of the Institute of Plant Industry. Vavilov was a pupil of Bateson. He was deeply versed in western science and western culture. He was sympathetic to the revolution but, like Darwin, he was of well-to-do origin and was, in any case, a scientist rather than a Marxist. The importance of the distinction will appear shortly.

During the following twelve years the growth of Vavilov's work was one of the most impressive signs of the scientific prosperity of the Soviet Union. Under his control experiment stations sprang up in European Russia, Transcaucasia, Central Asia, and the Far East. Under his leadership expeditions went to Afghanistan, Abyssinia, Mexico, and Peru. Today his pioneer exploration of the sources of cultivated plants is being studied and his example followed, both for its practical and for its theoretical value, by Dutch, American, and British scientists. In Soviet Russia, however, events took a different course. Under the new Stalin régime, after 1928, Marxist orthodoxy became more important. Party members of research institutes began to find it profitable to intrigue for promotion. Philosophical exegesis began to intrude more frequently into the papers of young scientists. Western influence became suspect. A screen was gradually drawn between Russia and the outside world. English summaries were reduced in scientific papers. Visits abroad became restricted to those who left hostages behind. Too many of the leading Soviet scientists, geneticists, as well as others, who went abroad were hesitating to return.

Gradually the new conditions began to have their effect. More

[2] Brother of the present President of the Soviet Academy of Sciences.

and more was Soviet science preached as a thing by itself, derived from a canon which included Marx, Engels, and Lenin and such as these had spoken well of. Marx had spoken well of Darwin so that Darwinism was sacred. Lenin had spoken well of Timiriazev so that his views were sacred in the second order. Timiriazev had spoken well of the American commercial plant breeder, Burbank, so that he enjoyed third-order sanctity. And, again, Lenin had spoken well of an aged Russian plant breeder named Michurin, of whom we shall hear more later.

The application of these apostolic methods began to have more effect as the authority of western science waned and the authority of Moscow philosophy, so-called Marxism, waxed and flourished. A Government which relied on the absence of inborn class and race differences in man as the basis of its political theory was naturally unhappy about a science of genetics which relies on the presence of such differences amongst plants and animals as the basis of evolution and of crop and stock improvement.[3] It was desirable to have a theory of genetics interpreted and controlled by Moscow. It was only necessary to go back to the Lamarckian notions that Darwin himself had dallied with and to say that these were Darwinism or at least 'Soviet Darwinism.' It could then be assumed that better food would breed better wheat and likewise better men. The idea was to be found in the folklore of all nations since the ancient Hebrews. It was an idea that had commended itself to Timiriazev in his dotage and to Michurin even in his prime. Their names began to be more and more quoted, and suitably selected passages from their writings more and more frequently published.

With these developments in mind it is worth while looking back on the histories of Michurin, and his American counterpart, Burbank. Both have enjoyed a great name in the popular esteem of their own countries, a name to which statesmen have paid proper posthumous regard. Roosevelt honoured Burbank with a purple postage stamp. Stalin paid his homage to Michurin with a small town, Michurinsk. In both countries societies were formed to advance their work or reputation. Both men had worked for their private profit, collecting useful plants from other countries and breeding from them. Their methods did not include such precautions as are taken by scientific plant breeders, but sometimes, as in all botanical collections, useful seedlings turned up by chance from the seed set by open pollination. In these cases both

[3] Consistency is, however, no longer sought in this respect. According to Ashby, the Soviet school curriculum on Darwinism includes 'The inadmissibility of extending the theory of natural selection to human society.' Such an extension is classified under heresies as 'social Darwinism.'

Michurin and Burbank felt able (as commercial breeders usually do) to attribute the results to 'scientific' crossing with particular, and often surprising, parents that happened to be growing nearby. Michurin claimed to have shifted the northward limit of cultivated fruits in Russia. But he made no mention of the fact that Sanders in Ottawa began his breeding work at the same time, in 1887, with the same object and with somewhat better authenticated success, using accepted scientific methods. Michurin merely admitted having received fruit trees from Canada and the United States.

In order to support their prodigious and, by scientific standards, fraudulent claims as creators of new plants, both Burbank and Michurin revived the good old Lamarckian theory of the direct action of a changed environment in changing heredity. They put the theory into a new dress and each probably thought he had invented it. But Michurin went further. He added a few ribbons that were entirely his own. He claimed that by grafting it is possible to *train* a plant for crossing, or to improve the quality of the result of crossing. Moreover, he could modulate or temper heredity by taking pollen from older or younger parents or by grafting older or younger seedlings. These principles a hundred generations of the most skilful and scientific hybridisers and grafters in other countries had failed, and have still failed, to discover.

Armed with such new and powerful instruments with which to change the course of nature, the great man's followers at Michurinsk successfully crossed apples with pears, plums with peaches, cherries with plums, and red with black currants. In other countries, if such crosses could be made at all, the hybrids would be intermediate between the two parents and completely sterile. The fact that Michurin's 'hybrids' were purely maternal and entirely fertile, squares perfectly with the evidence of his technical incompetence. They are not hybrids at all. They are, however, good enough material for propaganda when some of the particulars are omitted. The process by which an improbable statement of fact comes to be supported by a still more improbable theory which in turn comes to be supported by impossible statements of fact suggests an analogy with the development of those old religious myths which have fortunately been uprooted in Soviet Russia.

In view of these great successes of a more Marxist science, the situation of genetics in the Soviet Union as the first five year plan got under way was becoming what is called 'destabilised.' Scientific genetics was still officially taught and applied, under Vavilov, to plant and animal breeding, and, under Levit, to medical science. But a strong undercurrent of misgiving was making itself felt in political circles, an undercurrent the possibilities of which were well understood in the more remote and backward provincial universities and research stations.

It was in this situation that a dramatic political change made itself felt. The rise of Hitler to power gave new life to the forces working against western science in general and against genetics in particular. Hitler's doctrine was founded on giving a distorted predominance to a distorted genetics. His theory assumed the permanent, and unconditional, and homogeneous, genetic superiority of a particular group of people, those speaking his own language. The easy retort was obviously to repudiate genetics and put in its place a genuine Russian, proletarian, and if possible Marxist, science. For this purpose very little research was necessary: the classical personalities and achievements of Timiriazev and Michurin were there ready to hand. All that was needed was to discover a new prophet of Marxist genetics or Soviet Darwinism. The prophet was found in Trofim Dennissovitch Lysenko.

The first appearance of Lysenko was in 1928 in connection with special temperature treatments of wheat seed and plants by means of which the plant would come into ear earlier and could therefore be grown farther north. The discovery was based on the work of Gassner in Germany. It was Lysenko, however, a previously unknown worker at an agricultural research station in the Ukraine, who suggested the exploitation of this 'vernalisation' in Russia. His modest proposals were received with such willing faith that he found himself carried along on the crest of a wave of disciplined enthusiasm, a wave of such a magnitude as only totalitarian machinery can propagate. The whole world was overwhelmed by its success. Even Lysenko must have been surprised at an achievement which gave him an eminence shared only by the Dnieper Dam.

Perhaps Lysenko also became aware that such an eminence carried with it some vulnerability. For the area of land under vernalised wheat did not grow with the bubble of his reputation. Indeed, at the present day it is a moot point whether true vernalisation of cereals is any longer practised in Soviet Russia. The name merely persists as a reminder of past successes in the title of a journal still published by Lysenko.

In these anxious circumstances Lysenko seems to have concluded that the best method of defence was attack, and the people to attack were the geneticists who had been placed in a still more vulnerable position than himself by the rise of Hitler. For this purpose he was lucky enough to fall in with a most astute ally at Odessa; this was a philosophical, that is to say a Marxist, writer of the name of Prezent. The two together apparently took in hand the destruction of the Vavilov school.

The campaign against genetics was built up at a series of national genetic congresses which were held in 1932, 1936, and 1939. The first was mild and hardly noticeable. It was in fact premature. For the second the ground was more thoroughly tilled. Propaganda at all levels from the daily press to the scientific journal, and in all directions, practical and theoretical, political and philosophical, was brought to bear on genetics. A large and popular audience, to the number of 3,000, was marshalled in the conference hall. With this planned organisation Lysenko and his manager launched their attack. They presented their arguments on a correct philosophical and canonical basis which made experiment unnecessary; which was fortunate, for the experiments adduced were without controls, without definitions, and without numbers. In a word, they had no scientific meaning. Before a meeting suitably packed with party men these shortcomings proved to be no disadvantage. H. J. Muller,[4] the leading foreign exponent of the philosophy of the science, replied to the charges that had been brought against genetics, but the official report omitted his remarks. The Lysenko-Prezent program in 1936 was an almost entire success. At the end the chairman was fully convinced, and a resolution was passed that in future genetics and plant breeding were to conform with dialectical materialism. In other words, Moscow was to decide what was right and what was wrong.

Only one more meeting was needed to complete the work. This was held in 1939. It took the form less of a conference than a trial. The tone of the attackers became more aggressive and more authoritarian. Darwin himself said it. Timiriazev himself said it. Michurin himself said it. Such was the refrain. But pious quotation was no longer enough. Genetics, or 'mendelism-morganism' as it had now become, was connected with fascism and therefore with treason. Clearly no hope was left for the geneticists. Many confessed their errors. But Vavilov could hardly escape the evidence that he had been nominated President of the International Genetics Congress of 1939 in Edinburgh. The Commissar for Agriculture declared for Lysenko in practice and theory. The following year, Vavilov was dismissed from his executive post and arrested, while on duty in Rumania. Later he was condemned to death (on the ground, as it happens, of espionage for Great Britain). On July 31, 1941, he was taken out of one of the solitary condemned cells in the Butyrki prison and put in a larger cell

[4] Awarded the Nobel Prize for Medicine and Physiology in 1946.

with nineteen other political prisoners, whether on account of lack of space, or with a view to exile in Siberia, is unknown. When the Moscow prisons were evacuated in December, 1941, he was removed in distressing conditions to a concentration camp near Saratov. There he died.

In Vavilov's place Lysenko was appointed President of the Lenin Academy of Agricultural Sciences and Director of the Laboratory of Genetics (the name still stuck to it) of the Academy of Sciences. He also became a Vice-President of the Supreme Soviet. At the same time the industrious Prezent was rewarded with the chair of 'Darwinism' in the University of Leningrad.

This controversy differs in two respects from any other that has been known in modern times. When Roosevelt made the mistake over Burbank, the United States Department of Agriculture, the National Research Council, and the Carnegie Institution of Washington did not dismiss their very able staffs of geneticists. In Russia, however, Stalin's mistake was part of a plan which included the dismissal of the geneticists and a great deal more besides.

The first victims were in 1932 when G. A. Levitzky the cytologist and his pupil N. P. Avdoulov, were sent to labour camps. Vavilov asked Stalin for their release, a request which was temporarily and grudgingly granted. At the same time B. S. Chetverikov the pioneer of population genetics and W. P. Efroimson were sent to Siberia and nothing was heard of them for fifteen years. In 1935 the first two geneticists I. J. Agol and L. P. Ferry were put to death. In 1937 the head of the great institute for medical genetic research in Moscow, S. G. Levit, was put to death and at the same time probably Avdoulov. In 1939 N. A. Iljin, an outstanding animal breeder, disappeared, leaving a posthumous paper to appear in the *Journal of Genetics*. In 1942, not only Vavilov, but also his closest cytological colleague G. D. Karpechenko died ('in the fighting' so it was said). At about this time J. J. Kerkis the *Drosophila* geneticist disappeared following a party cell intrigue and Levitzky was finally sent back to a labour camp where he died. At about this time also N. K. Koltzov, the cytologist and doyen of Russian biology, died and his widow committed suicide.

In a word, after thirteen years of persecution, the great fellowship of Russian biological research, formed in the revolution, had been crushed and broken.

The details of this story are covered by secrecy mitigated by rumours and occasional apologies. Enemy action, currency speculation, political unreliability, fascist conspiracy and even, strangest of all, mere nature, have been alleged as causes of death. Presumably Vavilov did not live to know that he had been elected a Foreign Member of the Royal Society in 1942. Those who ask why it was that in the closing stages of the genetics trial he and his devoted followers seemed to have lost their nerve should consider that these men, knowing the methods of the Soviet Government, could foresee their own destruction.

The western scientist is bound to seek a parallel for these events in his own history. For men who died because they asserted the freedom of scientific inquiry from dogmatic control, nazism and fascism do not offer comparable examples. In Germany and Italy a sound genetics continued as an underground movement under fascism and has survived to-day. Even if we look back as far as Giordano Bruno, who died in 1600 at the hands of the Roman Inquisition, we find only a solitary victim. Never before has science been offered so many martyrs to its cause, men, too, honoured and beloved throughout the world. We must, however, follow the plot further.

Established in his new offices Lysenko was able to develop the theory and the practice of his Marxist 'Genetics.' The first he described in a small book called *Heredity and its Variability,* published in the stress of war in 1943, and circulated abroad in 1945 and again in 1946. It has now been translated—so far as writing of so transcendental a character can be translated—by one of the three most distinguished Russian geneticists in exile, Professor Th. Dobzhansky, of Columbia University.

Lysenko begins, on page 1, with a definition. Heredity, he says, is not what foreign genetics tries to make us believe, the property by which like begets like. On the contrary heredity is nature; and nature is development. We can, therefore, wipe out the misguided discoveries of bourgeois genetics. They are flat, stale and unprofitable. We can take it as proved without further evidence that, since changes in the environment directly modify the development of a plant or animal, they can also directly modify heredity; that is to say, change the character of all the descendants of a plant or animal which has been specially treated.

This discovery (which is in keeping with the edited words of the masters already referred to and presumably also with the resolution of 1936) is obviously of great value to the plant breeder. Theory and practice are immediately united. Lysenko's anxieties about the practical failure of vernalisation are removed. He can vernalise his wheat once and all succeeding generations will be born ready vernalised. With the wheat 'varieties' common in primitive cultivation, consisting as they do, not of one, but of great numbers of pure lines mixed together, selection can quickly sort out types adapted to new and extreme conditions. In this way Lysenko is able to claim that 'natural' vernalisation has altered the heredity of wheat grown in the far north and thus turn a physiological defeat, the failure of his vernalisation programme, into a genetic victory, the transformation of Russian agriculture. On nineteen out of the sixty-two pages of his work on heredity he adumbrates or reflects upon the expectation or promise of this practical achievement. In different countries there are different ways of keeping the wolf from the door.

The essential fallacy in Lysenko's, as in all would-be Lamarckian, experiments depends on starting with a mixture. Lysenko, however, denies the existence of pure lines and assumes that seed samples of all crops are mixtures; he may therefore count himself well protected against this charge. For a few years longer he can continue to hold the carrot of success in front of the donkey's nose and for a few years longer the vicious circle in which the pseudo-chemist, the politician and the philosopher deceive and bribe one another can follow its course.

So much for the serious side of Lysenko's book. There is, however, a comic side. A science must have its dialectical relationships. Lysenko (with the help perhaps of the new Professor of Darwinism) is quite willing to provide them. In order to do so he has to take elements of genuine genetics as his examples. Male and female germ cells do not carry permanent hereditary materials like the chromosomes of western genetics. On the contrary, in fertilisation, they digest one another. This has not been seen under the microscope, and what it would look like, nobody knows. The great Russian cytologists in any case are all dead. But it follows directly from the Marxist principle of the interpenetration of opposites. The stronger thus destroys the weaker for the time being. Hence 'dominance' is explained, although it may seem to us to be less of the kind described by Mendel than that referred to in an unfortunate passage of his fourth book (line 1,209) by Lucretius, when he suggests that children take after the more vigorous partner in the act which begets them. Since 1943 a further advance in theory has been made. Segregation is due to indigestion. The dominant gene (if one may use that bourgeois expression) with rustic vigour belches the recessive one. This negation of the negation is also strict dialectical materialism as now understood, and does not require vulgar empirical support.

How are we to describe this science or this philosophy? Anthropomorphism, teleology, animism, and necromancy (which Lysenko modestly calls 'Michurinism') are no doubt all dialectically reconcilable with current dogma in Moscow. To the uninitiated, however, their happiest analogy is to be found in the sacred writings and mystical ideas of Hinduism, and when Lysenko describes development as due to the 'unwinding of a spring wound up in the preceding generation' the Pandits of Benares might well recognize the doctrine of Karma and applaud so correct a philosopher.

In the career of Lysenko there is much of interest for men of science. Some things of course must be in doubt. It may be difficult in such a man to distinguish between the enthusiasm of the charlatan and the frenzy of the fanatic. It may seem that the denial of mysticism must itself become an act of mysticism in the mouths of those whose reason does not itself consent to the denial. It certainly will seem that dialectical processes are as unpredictable in their results as any other political manoeuvres. Unity of theory and practice may be achieved (to use the jargon of a decadent science) by the digestion of a recessive theory by a dominant practice and of both by an epistatic politics.

But there can be no doubt that the genetics controversy is in one respect much more serious than any of ours. It means the collapse of an important part of the scientific foundations of an improved agriculture on which the peoples of the Soviet Union depend for their food. Many millions of the more efficient and prosperous peasants have already perished in order to make way for communist improvements in Russian agriculture, that is, for what hopeful and kindly people in foreign parts supposed were scientific improvements. There are few countries in the world in which there is more room for such changes, especially by plant and animal breeding. There is now no ground for hope that such original expectations of Lenin's revolution are going to be fulfilled. And, as for the dialectical materialism in whose name certain branches of science are being crushed, it is neither dialectical nor materialism. It is humbug, humbug fostered and sheltered by a xenophobia, born of scientific obscurantism and political expediency. Lysenko and his followers are ignorant of foreign languages, and no foreign book on genetics or cytology has been translated into Russian since Stalin took control. Their notions of western genetics and cytology are therefore always twenty-five years out of date. Yet the bogus achievements of Lysenko and Michurin continue to be foisted on us (and still more aggressively on the Russian-occupied zone of Europe) by the Soviet propaganda machine. The window-dresser and the shop-gazer have ceased to be aware of the deception they are propagating, of the dream world they are living in.

The philosophical part of this deception is not the least instructive. Since the early statements of Marxist philosophy great scientific developments have taken place, most of them outside

Russia. These developments, particularly in genetics, give a strength and coherence to dialectical materialism which it did not originally possess. Marx knew no more of heredity than Darwin did. Perhaps a little less. But, while Darwin admitted he knew nothing, Marx has to be supposed by Marxists to have known, or to have foreseen, everything. The scholastic method imposed by Lenin, and still more by his successors, thus unfortunately stabilised the official and imperfect doctrine of Moscow in opposition to the highly materialistic and highly dialectical theory of heredity which has been developed since the time of Darwin and Marx in western countries, developed by clerics like Mendel and Janssens, and bourgeois intellectuals like Bateson and Morgan, obeying and using traditional scientific methods.

Why should Marxist philosophy, which was supposed to be founded on science, and to be permanently concordant with it, adopt a scholastic method which was bound to conflict with the later findings of science? The reason for this was that the materialist interpretation of history, which was politically fundamental for the Marxist in Russia, obviously could not be recast from time to time by scientists. Marxism had to be governed by politicians according to their own ideology if it was, like the papal system, to establish the politicians in control. To Marx, heredity was not part of the materialist interpretation because it was immaterial. When science put heredity on a material basis—and with it the basis of class distinction—Marxism was already petrified. We can see therefore how fatal an uncontrolled development of science might be to a political system ostensibly founded on science. To use the Marxist phrase, we can see how the internal contradiction developed in political Marxism.

It is a rule that proselytising religious communities and other missionary bodies forswear, to some extent, the principles and practices on which they have based their claims to political power when in the end they achieve that power. Never has this happened so rapidly as in Soviet Russia. Never certainly has there been so ironical a situation as that in which the authorities in Moscow reproach the rest of the world for an idealism and superstition which are everywhere being supplanted by a materialistic science except in Moscow—and perhaps Madrid.

Once science has been brought under control the change is hardly to be remedied. The further Soviet biology moves away from the natural materialistic development of science the more loudly will its prophets declaim against the wickedness of capitalist or fascist science and the more loudly will they proclaim the mystical merits of Soviet Darwinism or Michurinism. There are, to be sure, several aspirants waiting in Russia to pull Lysenko off his perch by the reintroduction of scientific method. The position of Vice-President of the Supreme Soviet may be envied even if it does not seem enviable to us. Doubtless someone will succeed. But he will not reverse the resolution of 1936. Nor will he bring back the dead to life.

Another question now arises: What bearing has this history on the fate of scientific research in general in the Soviet Union? From time to time other branches of science, less notable in Russia than genetics was, are cleaned up, and the survivors, physicists or psychologists, made to understand the correct lines of development until the next change of front. But apart from the caprices of a despotic Government there is something else in the Stalin system equally, although more gradually, fatal to science. The general requirements of secrecy and seclusion would hamstring the scientific research of any country in the long run. In Russia, which has never been above the fifth place in fundamental research, their effect is bound to be more rapid than it would be in any of the first four. But the most serious of all obstacles to Soviet science is the political control of the individual research worker, and of his privileges and promotion, at every stage of his education and training.

Science proceeds very largely by disagreements, controversies, or conflicts, processes which are resolved outside Russia by new dialectical transformations, new syntheses. Inside Russia they are resolved in the scientific, as in the political, field by the destruction of one of the parties to the conflict. The make-believe of free controversy, dressed up as for example in the Genetics Conferences, is pure pantomime. The only difficulty is to understand who is expected to enjoy the illusion. The Marx-Engels Institute in Moscow would be well employed in pointing out defects of this kind in the Soviet system, if such criticism were permissible. All these factors quite as much as the dramatic open plots, trials and persecutions which have led to the suppression of genetics, are bound to crush fundamental research in Russia into irretrievable ruin. The survivors of the pre-revolutionary generation are still in charge. When that diminishing company has gone we shall see the results of subjecting science to philosophy and philosophy to politics in a totalitarian hierarchy.

Many false conclusions can be, and no doubt will be, drawn from this history of Soviet science. The principles of the planning of research and its application to social welfare which are officially accepted in Soviet Russia must be applauded by men of science everywhere. We must also acknowledge that these principles are to a limited extent put into practice by the Soviet Government. But we cannot fail to see that three primary evils, by no means necessarily connected with either planning or socialist organization, but undoubtedly connected with one-party government, frustrate the hopes of a scientific civilisation in Soviet Russia. The first is the establishment of an officially interpreted orthodox philosophy. The second is the suppression of free speech and of free cultural relations with foreign countries. The third is the use of the death penalty against scientists, and against hostages taken from their families, to enforce this suppression. From these primary evils political intrigue and delation, the pretense of infallibility, the bogus philosophy, and the sweltering authoritarianism, are all derived. No amount of glorification of science, or of what passes for science, will compensate for these evils either in making the good opinion of western nations or, what is more important, in making the happiness and prosperity of the peoples who are ruled from Moscow.

2

N. I. VAVILOV, A MARTYR OF GENETICS[5] 1887-1942

TH. DOBZHANSKY

Biologists everywhere have been concerned for almost a decade about the fate of N. I. Vavilov, the most eminent of Russian geneticists. The magnitude of the tragedy of this man has, however, become known only recently. It is now clear that genetics owes homage to his memory not only because his contributions as an original investigator have been great and his work as organizer and leader of research has been outstanding. He has suffered martyrdom for genetics. In our time, being a scientist is not usually considered a dangerous occupation, but rather one of the ways to relative security and to the little amenities of middle class existence. Not so with Vavilov. He ignored personal comforts and gave himself up entirely to his work. In his last years he had to suffer the anguish of seeing the results of his efforts being pulled down by incompetents. He met death a prisoner on the bleak and forbidding shores of Eastern Siberia.

Nikolai Ivanovich Vavilov was born in 1887, a son of a very

[5] Reprinted by permission from the *Journal of Heredity* 38:227-32. 1947.

wealthy merchant. He came from the curious class of "illustrious merchantry" of old Russia, which for centuries tended to become a closed caste, expert at profit taking but not otherwise noted for cultural achievement. Yet, in the late nineteenth and early twentieth centuries this class suddenly produced a large crop of leaders in almost all fields of intellectual endeavor. Among these leaders were Nikolai Vavilov, the biologist, and his brother Sergei, a physicist and currently president of the Academy of Sciences of USSR. The young Vavilovs had the opportunity to receive the best education which money could purchase. Nikolai chose as his specialty biology and agriculture. He graduated from the Agricultural Academy at Petrovsko-Razumovskoe (near Moscow), and soon went to continue his education and research at Cambridge University in England. There he became a student and a close friend of William Bateson, and one of the group of pioneer geneticists of that day. In 1913 and 1914 he worked at the John Innes Horticultural Institute founded and organized by Bateson.

Even before his sojourn in England, Vavilov had begun to study the immunity of cereals to fungus diseases, and published several papers on this theoretically interesting and practically important subject (1913). He then proceeded to investigate the genetic basis of the immunity, and found it adequately describable in fairly simple Mendelian terms. After his return to Russia at the start of the first world war, Vavilov commenced his great work on the origin of cultivated plants. This monumental study, unquestionably Vavilov's most important research contribution, was published in 1926, in a book which remains a classic, namely *The Centers of Origin of Cultivated Plants*. Here Vavilov gave a synthesis of the mass of information accumulated in the botanical literature since de Candolle, and developed his theory of several principal centers of origin and of concentration of genetic diversity in cultivated plant species. This work brought him world-wide recognition and acclaim. His other important theory, that of "The Law of Homologous Series in Variation" (1920, 1922, 1935), is an empirical generalization which states that related biological species tend to parallel each other in hereditary variability.

ORGANIZATION OF SOVIET GENETICS

The Russian revolution opened an unlimited field of application for Vavilov's energies. After an interlude as a professor at the University of Saratov (1917-1920), he was placed in charge of the Bureau of Applied Botany in Petrograd (1920). Within a few years, he developed this rather inconsequential bureau of the old Ministry of Agriculture into one of the largest and most active research institutions devoted to agricultural sciences anywhere in the world. He not only did a prodigious amount of organizing work at home but also traveled all over Europe and the United States to establish contacts with scientists in other countries. Most important, he brought back wih him a whole library of scientific literature, the supply of which had been cut off for several years by the war and, afterward, by the blockade of the revolution. Active biologists from all over Russia flocked to the library of Vavilov's new Institute to become acquainted with the advances of world science. Not a few of these visitors ended by joining Vavilov's rapidly growing staff of collaborators. Others undertook new types of research integrated with the work of the Vavilov group. Still others carried away with them some of the inspiration and enthusiasm which prevailed in Vavilov's entourage. In a few years, the All-Union Institute of Applied Botany and New Crops, as the Bureau was now called, became the center of a federation of agricultural research institutes distributed all over the USSR, from the Polar Circle to the subtropics of Caucasus and Turkestan. The combined staffs of these institutes in 1934 amounted to about 20,000 persons.

Despite the enormous weight of his administrative responsibilities, Vavilov managed to remain the *de facto* scientific and research leader, as well as the president, of this unprecedented research organization, which underwent one more change of name, to become known finally as Lenin's All-Union Academy of Agricultural Sciences. His seemingly boundless energy and vitality sufficed both for administrative duties and for creative research. Recognition came to him from all sides. He was made a member of the Soviet's Central Executive Committee (corresponding approximately to the American Senate), and was one of the very few non-communist members of that body. In 1929, he was elected member of the Academy of Sciences of USSR, and in 1931 president of the All-Union Geographical Society and director of the Genetics Institute of the Academy.

WORLD TRAVELER

The practical problems of the improvement of agriculture in USSR were approached by Vavilov with his characteristically sweeping breadth and vision. He recognized that new and improved varieties of economic plants can be created only by combining together valuable genes. Such genes can be found scattered in the existing varieties, thoroughbred as well as primitive ones, all over the world and, especially, at the "Centers of Origin." A scientific breeder, he realized, must first of all take stock of the available resources, of the genetic raw materials, which will be useful to him in his work. Such a global inventory is completely out of the question for an individual scientist, or for a small group of scientists. But an Institute such as he had built, backed by the resources of a socialist state, could, so thought Vavilov, undertake this historic assignment. Hence, Vavilov and his colleagues journeyed and collected scientific materials in approximately sixty countries in all parts of the world. The scientific and practical value of the collections assembled in the institutions administered by Vavilov was inestimable.

Vavilov himself was probably the most widely traveled biologist of our day. He visited the United States and Western Europe several times, and was personally well known to, and liked by, most of his genetical and botanical contemporaries. But he was also an explorer not afraid to penetrate, with a very modest, not to say meager, financial support, places remote from well-trodden paths. As early as 1916, he had explored Persia, Turkomania and Bokhara. In 1920-1923 he visited various parts of Middle Asia, including Tadjikstan and Pamir. In 1924 he covered a long and arduous itinerary in Afghanistan. In 1925 he went to Khiva. In 1927 he studied the Mediterranean region from Portugal and Morocco to Syria and Trans-Jordan, and made an expedition to Ethiopia and Somaliland. In 1929 he visited Chinese Turkestan (Tarim Depression) and Songaria. Finally, in 1930 and 1932-1933 he traveled and collected extensively in Mexico and in Central and South America. The country which he was unable to visit despite his eagerness to do so was India, because its government steadfastly refused to admit him. Professor H. J. Muller informs me that, by a tragic irony, Vavilov was finally invited to visit India in the fall or winter of 1937. But when this opportunity came at last, he was refused permission to go by his own government, that of the USSR. The results of his activity can be gleaned from the fact that more that 25,000 samples of wheats alone were collected and grown in experimental plantings in various parts of USSR.

DEVOTION TO SCIENCE

Vavilov was first and foremost a man of action. His energy, forcefulness, and working ability were marvelous. He was actually able to get along on between four and six hours of sleep per day, and appeared to neither need nor desire any rest or recreation. No wonder that his collaborators considered it as something less than a privilege to travel or to live in his company for many days in succession. Comforts and conveniences counted for little with Vavilov. In the nineteen-twenties, while he was one of the most influential scientists in the USSR, he lived in his office at the Institute of Applied Botany, an oldish leather-covered davenport serving as his bed. His meals were prepared by a janitor's wife, who did not excel in the art of cooking. When he married his scientific collaborator, E. I. Barulina, the new Mrs. Vavilov discovered that the household expenses for the first few months had to be met chiefly from her own salary. His much larger salary had been pledged for assistance to all manner of people, some of them only slightly known to him.

Despite being forced to lead the life of a busy executive, Vavilov neither abandoned his own creative research nor let it be performed in his name by assistants. More than that, he found time to read the current scientific literature, and was always fully conversant with the recent advances in genetics and in agricultural research, which is something very few scientists holding executive positions manage to do. He always enjoyed discussing current research problems with other scientists, even with beginners. But to have the privilege of such a discussion, one was often given an appointment at some altogether unorthodox hour. Vavilov did not have one iota of the self-conscious eccentricity of an eminent man, and his manner of address was equally direct, cheerful, and sincere with his equals and with his juniors.

To Vavilov, all worth-while scientific problems seemed to have bearing on the welfare of the whole world, and, hence, the whole world had to be called on to contribute toward their solution. Almost every publication that came from his pen attests this truly cosmopolitan spirit of its author. He was irresistibly fascinated by grand scales and by the world-wide implications of his ideas. In the words of one of his friends, a fundamental trait of Vavilov's personality was that he disliked quantities of less than a million. And, yet, Vavilov was an ardent Russian patriot. Outside of Russia, he was regarded by some as a communist, which he was not. But he did wholeheartedly accept the revolution, because he believed that it opened broader possibilities for the development of the land and of the people of Russia than would have been otherwise. In October 1930, during a trip to the Sequoia National Park in the company of this writer (and with nobody else present), he said with much emphasis and conviction that, in his opinion, the opportunities for serving mankind which existed in the USSR were so great and so inspiring that for their sake one must learn to overlook the cruelties of the regime. He asserted that nowhere else in the world was the work of scientists appreciated more than in the USSR.

THE LYSENKO CONTROVERSY

A Congress of Genetics, Plant and Animal Breeding, attended by about 1400 members, assembled in Leningrad in 1929 under the presidency of Vavilov. Among 348 papers read at this Congress, a fairly interesting but in no way revolutionary study on the physiology of cereals had as its junior author one T. D. Lysenko. A few years later the name of Lysenko was destined to become familiar not only to biologists but to newspaper readers throughout the USSR. He was hailed as the discoverer of vernalization, a process whereby winter wheat can be influenced to produce a crop if sown in the spring. The phenomenon of vernalization had been discovered in the United States years before Lysenko gave it a name; but Lysenko certainly proved himself a master of the art of modern publicity. He claimed, or it was claimed for him, that vernalization inaugurated a new era in Soviet agriculture, permitting, among other things, the culture of cereal crops much farther north than was formerly possible. The vernalization bandwagon was highly popular some ten to fifteen years ago, but it is perhaps significant that little has been heard about practical applications of vernalization in the USSR or anywhere else in recent years.

Vavilov welcomed Lysenko's debut; although his published praises of Lysenko sound a bit hollow, he urged facilities for testing Lysenko's ideas. Lysenko was, however, interested in much bigger stakes. Sometime in the early nineteen-thirties, Lysenko formed an alliance with I. I. Prezent. Prezent was neither a biologist nor an agriculturist, but a specialist in the philosophy of dialectical materialism; he was also a highly effective polemical speaker and writer, and a possessor of a cultural refinement conspicuously lacking in Lysenko. In 1935 and 1936, Lysenko, Prezent, and their followers struck. In a stream of magazine and newspaper articles and speeches, they declared genetics to be inconsistent with dialectical materialism and with Darwinism as they construed the latter, and to be, in fact, tainted with fascism and with Nazi race theories. Furthermore, they contended, Vavilov's basing the work of plant and animal improvement on genetic principles had caused inexcusable delays in the successful outcome of this improvement work. Vastly more spectacular, and anyway vastly more rapid, practical attainments would come if only Vavilov's mismanagement and the suspect "Mendelian-Morganian" genetics were supplanted by Lysenko's patriotic leadership and the incorruptibly dialectico-Darwinistic approach.

In 1935 it was a deadly serious matter to be accused of having slowed down the development of agricultural production in

USSR. To consider these charges a new Congress on genetics and agriculture was convened in Moscow in 1936, presided over by A. I. Muralov, a high governmental dignitary. The published transactions of this Congress make painful reading. There were Lysenko and Prezent with a well-organized group of followers, pleading, cajoling, and threatening. Several geneticists, among them H. J. Muller, the visiting American, vainly tried to stem the tide against genetics. The least inspiring sight was that of some competent scientists who attempted to sit on the fence or who made unctuous speeches praising both factions. Vavilov himself made two speeches in defense of modern genetics and agricultural science. To judge from the published texts, those speeches lacked Vavilov's customary forcefulness and optimism, as though he felt that the issue had already been decided against him. And indeed, the 1936 genetics Congress turned on the whole against Vavilov, just as the one held in 1929 gave him his greatest triumph.

This rejection of sound scientific principles of established practical value, in favor of a witchcraft supported only by artful propaganda and by big promises, seems utterly incomprehensible. The subversion and demolition of the work on plant improvement organized so successfully and on such a vast scale by Vavilov undoubtedly caused a setback in the development of agriculture in the USSR. Since, even with the energy of another Vavilov, such an organization could not be restored overnight, this blunder has harmed, and will continue to hamstring this development for some time to come.

No matter what else may be said about their intentions, those assembled at the 1936 Genetics Congress, including a majority of Lysenko's followers, doubtless sincerely desired the betterment rather than the deterioration of Soviet agriculture. This paradox can be understood only in connection with certain peculiar features of the development of biology, and particularly of evolutionary thought, in the USSR.

GENETICS AND MARXISM

Some of these features have recently been analyzed by Hudson and Richens and by Beale.[6] Without going into detail, it may be stated that Lysenko and Prezent have exploited for their own ends an old antagonism toward genetics which had existed amongst some biologists but also to a greater extent among the general reading public in USSR. This antagonism arose because

[6] P. S. Hudson and R. H. Richens, *The New Genetics in the Soviet Union,* Cambridge, 1946; G. H. Beale, "Timiriazev, Founder of Soviet Genetics," *Nature,* Vol. 159, 1947.

of an unfortunate misunderstanding of the meaning and implications of genetics by Timiriazev, a highly respected intellectual leader, and by Michurin, a successful horticulturist. Timiriazev and Michurin regarded the early work of genetics, especially that of Bateson's school, as subversive to Darwin's evolution theory and, in fact, a product of "clerical reaction" against evolutionary biology. The opinions of Timiriazev and Michurin carried, and still carry, great weight not only because of their scientific authority but also because of the political eminence they achieved. Vavilov and the other geneticists in the USSR were, of course, aware of this antagonism, but they hoped that it would be dissipated as a better understanding penetrated the public mind. Lysenko and Prezent fanned the antagonism to an inferno of contention in their attempt to unseat Vavilov and to grasp his place of leadership for themselves. Accusations of neglect of practical work were combined with indictments for heresy against approved philosophic principles.

In any scientific community individuals are not unknown who try to build their reputations by criticizing the work of others rather than by producing original ideas or work of their own. After the Moscow Genetics meeting in 1936, there was an open season for fault-finding regarding Vavilov's research and organizing activity. Dozens of hitherto unknown authors suddenly discovered that Vavilov's theories of the Centers of Origin and of homologous series in variation were totally unfounded. Worse than that, those theories had led Vavilov to dissipate his efforts by sending expeditions to many foreign lands, instead of confining himself to studies on local varieties in the USSR, which would have resulted in greater practical achievements. Vavilov was also accused of causing plant breeders to rely on the method of sexual hybridization which is practiced everywhere in the world, instead of the method of "vegetative hybrids" proposed by Michurin and Lysenko. Vavilov was told patronizingly by a certain I. M. Poliakov that "it is not necessary for you to bow slavishly before foreign science." And indeed, some breeders in the USSR promptly switched from sexual to "vegetative" hybridization. But Vavilov's worst sin, which negated all his research and practical activities, was his backsliding from canonical Darwinism (as handed down from Darwin through Timiriazev, Michurin, and Lysenko) into the Mendelian-Morganian heresy. This setting of Darwin as an incontrovertible authority in opposition to genetics, is one of the weirdest chapters in the unbelievable story of Lysenko's rise to power. The fact that genetics is the foundation of modern Darwinism proved to be no obstacle in the Lysenko-Prezent campaign.

DOWNFALL AND EXILE

In August 1939, the Seventh International Congress of Genetics was held at Edinburgh, Scotland, and Vavilov was invited to become its President, thus receiving the highest honor which the consensus of opinion of the world's geneticists can bestow. He accepted the invitation. But less than a month before the Congress was to open, came a letter, signed by Vavilov, which stated that "Soviet geneticists and plant and animal breeders do not consider it possible to take part in the Congress," because the latter was to be held outside the USSR. Few if any members of the Congress had any illusions as to whether Vavilov was a free agent when signing this letter. Matters were moving rapidly toward a denouement. In October 1939, a "Conference on Genetics and Selection" was held in Moscow, at which the problems thrashed out at the 1936 meeting were gone over again, with Lysenko, Prezent, and others greatly expanding their claims as to the theoretical soundness and practical efficacy of their "Darwinism." Vavilov, interrupted and heckled from the floor, delivered what was probably the weakest speech in his life, his attitude being almost entirely defensive, although he courageously reasserted the soundness of the basic principles of genetics. He evidently was already a broken man.

After the 1939 Genetics Conference, a shroud of silence envelops Vavilov. The closing chapter can be reconstructed only from unofficial, fragmentary, but apparently reliable information. Vavilov was arrested, probably in 1940. Part of the time during the winter 1941-1942, he was a prisoner in a concentration camp at Saratov (ironically, it was at the University of Saratov that he held his first post under the new revolutionary regime), and whence he was transported to Siberia. His destination was Magadan, on the Sea of Okhotsk, the capital of a rich gold-bearing region, but a place of sinister reputation, because of its deadly climate and even worse because it was built and operated by forced labor. According to some information, Vavilov was put to work on breeding varieties of vegetables capable of growing in Magadan's climate, but this information is not certain. The release, through death, probably came in late 1942. No mention of N. I. Vavilov's name can be found in the list of living and recently deceased members published by the Academy of Sciences of USSR in connection with its 220-year jubilee celebrated in 1945.

Vavilov's life was connected so intimately with the development of genetics in the USSR that his martyrdom is not separable from the crisis of the Russian branch of that science. The bold and comprehensive long-range program of improvement of cultivated plants inaugurated by Vavilov in the USSR was meant to take more years than were given to it. This program, except for its early fruits, is probably lost. Although the available information is still too incomplete to permit a clear view of the situation. Vavilov is certainly not the only geneticist who fell victim of the wrath of the self-styled "Darwinists." And yet, it is assuredly not true that all genetic research has been suppressed in USSR, as some writers in American journals hastily asserted. To be sure, Lysenko had the temerity to demand that the teaching of genetics be discontinued in all institutions of higher learning in the USSR, but fortunately his influence was never sufficiently great outside the sphere of agriculture. As the uninterrupted flow of publications demonstrates, enough first-rate genetic research is now being done to enable the USSR at least to retain one of the places of prominence which it has secured, in part owing to Vavilov's organizing and creative activity, among the nations of the world.

3

GENETICS IN RELATION TO MODERN SCIENCE[7]

An Excerpt from Section Three of the Presidential Address Delivered on July 7, 1948, at the Eighth International Congress of Genetics Held at Stockholm

H. J. MULLER

The following observations are set forth only as those of an individual who has happened to come in contact with these matters at first hand. In presenting them, I do not speak in the role of an official of this Congress, for I do not wish anyone else to be held responsible for my remarks. In fact no other Congress member has been informed of them in advance. Yet I am convinced that these things should be stated here.

Despite the remarkable growth of genetics and of other branches of science and culture in the USSR following the overthrow of the Tsarist oppression, the distribution of forces affecting the long-range development of science is better illustrated

[7] We are indebted to the editor of *Hereditas* for permission to publish this excerpt.

by the fact that, for example, the official head of science in 1937 in the governing party—one K. J. Bauman—had as his own special field what was called "political science." In each scientific as well as other institution there is the local political unit or cell, usually consisting of those who are weakest as scientists, to keep watch, to stir up, to engage in expedient defamations and glorifications, to redirect, push, or inhibit, and to pass the word along to those politically higher up. But the orders from the higher-up political units, once decided upon by the latter, are absolute for those below, as in an army.

In that country, unfortunately, it has come to pass that genetics, as we know it, no longer finds a place in the official curricula, and that it is regarded by the dominant group of officials as a dire heresy. Most of the chief geneticists are, as the case may be, broken, executed, disgraced, not to be located, fugitives, forced into other work or, in the mildest cases, driven to such a redirection of their lines of research and publication as to give the appearance of working somehow in support of the doctrines approved by the officials. These officials have not been educated in modern natural science as we know it, and they have failed to understand its modes of procedure. All honor to those workers in our science who have suffered in this cause: to the heroic and tragic names of Nicolai Ivanovitch Vavilov, Philipchenko, Chetverikov, Karpechenko, Levit, Agol, Kerkis, Efroimson, Ferry, Serebrovski, Levitski and many others whom we will leave unnamed.

Is it not obvious that the first prerequisite of healthy science is democracy of minds, of thought, and that this in turn requires in the long run a generalized kind of democracy, involving, among other things, universal education in the results and methods of objective science? It cannot be claimed that this ideal has as yet been attained anywhere. In the meantime, however, we must recognize that what is most directly antithetical to scientific development is any form of authoritarianism. And more especially is this antiscientific when it imposes, as in the USSR today, a fixed, all-embracing, and mystical creed, one old-fashioned in its design and full of shibboleths, that claims to underlie all other thought. Such a creed, under the imposing name of "dialectical materialism," is taught to and interpreted for hundreds of millions in the USSR and its outlying territories today, in such wise as to draw those conclusions that happen to suit the few who wield the ultimate control over science, education, and life in general. To be sure, certain expressly favored departments of science, and more especially of such branches as engineering or ballistics, on the one hand, or the study of number theory or of nebulae, on the other, may for a time continue to flourish under such circumstances. However, while such conditions last, the different fields will wither or become infertile in direct proportion to their nearness to concepts of general significance in present human living. And it happens that genetics does have great general significance of this kind.

The disastrous effect which the Lamarckianism of the official creed must be having through its application of unscientific principles in the breeding of cultivated plants and domestic animals is evident to all geneticists. What is not so well known is the dangerous interpretation of genetics in man to which this doctrine leads. To understand this, a few facts of the history of a dozen years ago may be recounted. Before it was decided to call off the Seventh International Genetics Congress that had been scheduled to be held in 1937 in the USSR, it had been proposed to allow it, provided all sessions and papers dealing with man were eliminated. At about the same time, the plans for a volume which was being prepared by leading geneticists of the world, designed to refute the Nazi racist doctrines, were abandoned. The clue was given when one of the men highest in the administration of applications of biological science, the head of agriculture in the Party, still asserted, even after having presided at most of the 1936 controversy on genetics, that evolution could not have occurred without the inheritance of acquired characteristics. On hearing this remark, the present writer asked the administrator whether this doctrine would not imply that the colonial, minority, and primitive peoples, those who had had less chance for mental and physical development, were not also genetically less advanced than the dominant ones. "Ah, yes," he replied in confidential manner, and after some hesitation, "yes, we must admit that this is after all true. They are in fact inferior to *us* biologically in every respect, including their heredity. And that," he added, "is in fact the *official* doctrine." The word "official" here was his, although the italics are ours. And at this point we may interject the question "Just who could be more official on this subject than this particular individual?" The answer to this seems clear. "But," continued our authority, "after two or three generations of living under conditions of Socialism, their *genes* would have so improved that *then* we would all be equal." (Italics again ours.)

Here the cat is out of the bag. The Nazis, according to this confession, would be in large part right after all, although not completely so. But "we," the upholders of the "official" doctrine, must not let the people, the masses, know about that, for it would cause them to doubt our sincerity, and so we must by all means avoid any airing of the subject and deny such statements if they are made. And then finally, after a few generations, the difficulty will have passed away, and genetics in man can be left to take care of itself.

It is unnecessary to tell this audience that responsible geneticists do not share the belief that adversity leads to genetic inferiority nor the complementary one that good conditions improve the genes, but it is important that outsiders to genetics should know our position on these points. It is evident that we are much more aware than the Russian authorities of the danger of judging genetic composition by outward characters when the environmental conditions are significantly different. We are therefore far more cognizant than they of the Nazis' criminal error in acting on the presumption that the level of social dominance of a human group is a measure of its genetic worth. We must marvel, however, at the touching faith that these heads of Russian science and agriculture have in the genes' becoming automatically improved, through the unexplained miracle of a generalized, ready-made somatic adaptability, which precedes the inherited predisposition—surely a perfecting principle which itself "just growed."

Coördinate with this positive mysticism is the negative one involved in the Russian officials' denunciation of "Malthusianism," despite the fact that this principle, taken in a broad sense, formed a cornerstone of Darwinism. Through this repudiation it can only be implied that the earth can be made to afford unlimited room for the indefinitely continued geometric increase of all its human inhabitants. Moreover, the basic necessity, dependent on Malthusian limitations, of a *differential* increase of genes, not only for the furtherance of genetic adaptation but even for its mere maintenance, must at the same time be wishfully overlooked in the interests of the official, escapist doctrine, self-styled "Darwinism." But even this doctrine must have, in this rapidly moving world, its own evolution, and though I am not conversant at first hand with its later developments, it is said that by this time the very existence of genes is denied, and that to use the word openly is now dangerous.[8]

[8] Approximately one month after the above words were spoken, the Central Committee of the Communist Party of the USSR announced its approval of the doctrine of inheritance of acquired characters, and of the attack on genetics by the charlatan Lysenko. Several geneticists known to be Party members thereupon recanted their science. Shortly afterwards the Praesidium of the Academy of Sciences of the USSR is reported to have discharged with dishonor the following eminent scientists for their adherence to genetics: the physiologist Orbeli, the morphogeneticist Schmalhausen, and the geneticist Dubinin, abolishing the latter's laboratory of cytogenetics. It is further reported that the Academy of Agricultural Sciences has called for the rewriting of all

I am well aware that to speak of these preposterous things is far from tactful. It does not turn the other cheek. It is not even courteous, and it certainly does not seem to promote the friendship so needed in these critical times. I was in fact advised by a very eminent liberal American politician to say in these connections only what would further international friendship. My answer can only be that *illusions are mankind's worst enemies,* and that friendships founded on illusions may in the end be more dangerous than thorough understanding of one another.

Science in general cannot be a species of diplomacy, though diplomacy may be a kind of science. And though scientists, being human, do have their directions of interest, their outlook, and their methods largely limited by their own social background and milieu, nevertheless they must try to develop their science, and its relation with the general scheme of things, as nearly in accordance as possible with principles that have been arrived at by the objective painstaking procedures known best to them. These procedures, this set of methods of science, constitute the greatest accomplishments of mankind in his struggle from barbarism towards an understanding of himself and the universe.

Scientists in thus discovering or contributing toward new principles of general import have in all ages risked the contumely or worse of officials, of the public, and even of their fellows, as so ably pointed out reecntly by Darlington in his brilliant lecture "The Conflict between Science and Society." For new findings upset cherished traditions and vested interests, and scientists themselves are seldom personally powerful. As one illustration I am reminded of the sad, disillusioned remark made to me by an eminent animal-geneticist of the USSR when he said that, contrary to his earlier opinion, certain obvious applications of genetics in man, involving artificial insemination, were not to be thought of for a thousand years. I for one refuse to be silent for a thousand years, however. And I believe that genetics has not yet, in Western countries, gone so far toward becoming an underground movement as to make such an eclipse necessary. Let us struggle to keep our science, in all its ramifications, connections, extensions, and even speculations, in the light of day.

---

textbooks and revision of courses in biology and related fields so as to remove all traces of genetics. These publicized announcements on the part of the authorities themselves make the situation of science in the USSR now unambiguous, and make it impossible to hold that the above remarks misrepresent the case. (Footnote added August 31, 1948.)

V

The Beginning of the End

1

AN INTRODUCTION TO LYSENKO AND HIS CLAIMS

That a simple charlatan can lead a successful attack on a science and extirpate it from a supposedly great nation is preposterous but not without precedent. Nor is it beyond the range of rational explanation. Since the war we have learned that science in Germany was greatly handicapped during the hostilities because German scientists were placed under the supreme authority of an ignoramus. Even Hitler himself was in the personal care of a quack physician who damaged him considerably, but by no means enough really to benefit civilization. We have innumerable records of medieval autocrats seeking the advice of court astrologers, and classical history contains many instances of affairs of state waiting upon the advice of soothsayers. So there is nothing essentially new in the Central Committee of the Communist Party accepting the fraudulent claims of a man who is unable to deceive a single reputable scientist.

Unfortunately the Communist authorities cannot be deceived equally easily in all scientific fields. An incompetent engineer need design but a single bridge to be exposed, and an ignorant chemist may reach a sudden end as the result of a single experiment. In medicine and agriculture, however, the detection of incompetence is not so simple. Indeed, in the light of our present knowledge, we can recognize certain periods in the history of medicine when the physician was a distinct hazard to the return of health, but his patients frequently recovered in spite of the regimen he prescribed. In agriculture also, quacks are not easily exposed and crops will grow in spite of the ministrations of cultists. Before an agricultural practice can be evaluated quantitatively it must be measured in carefully controlled experiments. Even when the greatest care is taken, an accurate interpretation of the results is not a simple matter. It is certainly beyond the capacities of untrained commissars.

We really should not expect the political leaders of a nation to be competent to decide matters involving scientific subtleties. Their talents lie elsewhere. The best we can hope for logically is that they pick the right scientific advisers or even allow their advisers to be chosen by free and reputable scientific organizations. Of course, in a country where the scientists dare not criticize their totalitarian rulers there can be no objective body of scientific opinion, and consequently there can be no adequate check on the actions of the scientifically illiterate. In such an atmosphere the oversimplifications of quackery tend to spread unchecked, and charlatans who are politically orthodox and operate in the complex fields of applied biology flourish as the green bay tree. They may even look forward to years of prosperity and safety if they can only destroy those who have the technical knowledge to expose them.

The claims made for Lysenko's agricultural improvements are stupendous. However, the more we endeavor to find out just what he has done, the more vague his accomplishments seem. His two books which have been translated into English, *Heredity and Its Variability* and *The State of Biology Today* (reprinted in this chapter as *The Situation in Biological Science*), contain no scientific data of any kind. His pretensions are vast but unsupported. Some of the innovations ascribed to him are either not his at all, e.g., vernalization (page 26) or they are archaic, e.g., graft hybrids. Most of his procedures have been tested at various times during the last hundred years and have been found wanting. They certainly do not impress anyone who is familiar with the history of botany.

We are, however, face to face with a situation which demands an explanation. Lysenko now controls Russian research in agriculture. Inasmuch as the total agricultural production of a country can be measured with fair accuracy, there should be nothing uncertain concerning the size of the crops in the land of the Soviets. The commissars should know whether or not the agricultural yield is increasing and whether cultivation is spreading into new territories. The Russians certainly do not fool themselves on this issue. What evidence we have indicates that their agriculture has made great strides. As they uniformly credit Lysenko with this improvement, the question arises: how can we dismiss him as an ignorant quack?

The answer is easy. Actually there is no real incompatibility between an advancing agriculture in Russia and Lysenko's antics.

When we remember that before his fall from power Vavilov introduced over twenty-five thousand foreign varieties of cultivated plants into Russia, we would hardly expect Russian agriculture to be stationary. Here indeed was a marvelous opportunity for any group who could seize control of agricultural research, if they could at the same time silence or intimidate the geneticists. Once in control, there are several possible procedures for them to follow. For example, the foreign varieties could be subjected to harmless treatments—treatments which would not alter them in any way—and in a year or two they could be presented safely to Russian farmers as new creations. Agricultural workers who were not trained scientists might even be deceived themselves and be led to believe that they had produced new races; actually some of the more stupid Lysenko's followers appear to be perfectly sincere in their claims. But even this type of chicanery, if this is what actually happened, is not original with the Michurinists, for unfortunately in the past the marketing of imported plants as newly created hybrids was not unknown outside of Russia.

Any geneticist reading the reports of the 1948 session of the Lenin Academy will have no difficulty in recognizing which members of the Lysenko coterie had intelligence enough to know what they were doing and which were simply out of their depth. Surprising as it may be to some, Lysenko belongs in the latter group. Compared with Mitin or Prezent he seems stupid. Perhaps his personal hostility to Mendelian inheritance is caused by what (the editor has been told) is known in Russia as a *nieposobni-komplex,* a sort of frustration complex. Unable to handle the simplest mathematics, Lysenko resents it violently and denounces any application of mathematics to a biological problem. This puts Mendelism beyond his reach. As he equates all genetics with the 3 to 1 ratio, it is evident that he comprehends practically nothing of the modern developments in this field, and his complex makes him resent the very existence of a science which frustrates him. Corroborating this view we have the following direct testimony to the effect that Lysenko is baffled by Mendelian genetics.

Dr. S. C. Harland is one of the few Western geneticists who have had personal contact with Lysenko. The next two paragraphs are taken from an address he gave over the British radio (from *The Listener,* December 9, 1948):

I was invited to visit the Soviet Union in 1933, and Vavilov took me on a tour that lasted nearly four months, not only to the main experiment stations of Russia proper, but also far into Central Asia to Tashkent and beyond that nearly to the Chinese frontier. In Odessa we went to see a young man named Trofim Lysenko, who Vavilov said was working on the vernalization or treatment of seeds in order to secure earlier maturity or greater productivity. I interviewed Lysenko for nearly three hours. I found him completely ignorant of the elementary principles of genetics and plant physiology. I myself have worked on genetics and plant-breeding for some thirty-five years, and I can quite honestly say that to talk to Lysenko was like trying to explain the differential calculus to a man who did not know his twelve times table. He was, in short, what I should call a biological circle squarer. Vavilov was more tolerant of Lysenko than I was. He said the factors of the environment had never been adequately studied by anybody; that we ought to explore the environment; and that young men like Lysenko who "walked by faith and not by sight" might even discover something. He might even discover how to grow bananas in Moscow. Finally Vavilov said that "Lysenko was an angry species," and that as all progress in the world had been made by angry men, let Lysenko go on working. It did no harm and might sometime do good.

We all know the sort of man who claims to have proved that some branch of science is completely wrong, or out of date, or to have discovered perpetual motion. Usually such people go about the world muttering and no harm is done. But when, as I think is clearly the case with Lysenko, a man puts forward a novel form of pseudoscience which promises miracles, combines it with almost hypnotic oratorical powers and preaches these pseudoscientific doctrines with great malevolence and fanatical ferocity, we can hardly wonder that the rulers of the Soviet Union succumb to his pretensions and bestow their official blessing upon him. So it was that the power of Lysenko increased and that of Vavilov waned. Vavilov was accused of holding Fascist views. The nasty witch hunt had begun.

2

THE SITUATION IN BIOLOGICAL SCIENCE

Report of Academician T. D. Lysenko at the Session of the V. I. Lenin Academy of Agricultural Science

[*This address was delivered on July 31, 1948, and was published in* Pravda *on August 4 and 5. The following translation should not be offered to the reader without an explanation, particularly as another English translation has been available for some time. Early this year the International Publishers of New*

York issued a book by Lysenko entitled The Science of Biology Today, *a book consisting of this report together with the concluding remarks made by Lysenko on August 7 (pages 249-61) as the session came to an end. The translation used by the International Publishers turns out to be official, for it is included in the English edition of the* Proceedings *of the Lenin Academy issued by the Foreign Language Publishing House of Moscow. Some readers might like to compare the official version with the one that follows. We wish to call attention to certain points of difference.*

1. The content of the official translations and ours are almost identical. There has been no official distortion of meaning.

2. The official translation is smoother and in somewhat better English than ours, but ours is more literal.

3. Scholars consulted by the editor have agreed that Lysenko expresses himself in very bad Russian. We have tried to keep Lysenko's flavor (this side of incoherence) if mediocre English can ever be given the flavor of bad Russian.

4. Our version contains certain crude expressions used by Lysenko which are not in the official translation.]

(1) BIOLOGICAL SCIENCE—BASIS OF AGRONOMY

Agronomic science deals with living things—with plants, with animals, and with microörganisms. Therefore a knowledge of biological mechanisms is included in the theoretical basis of agronomy. The more profoundly biology reveals the mechanisms of life and the development of living things, the more effective is agronomy.

Fundamentally agronomy is inseparable from biology. To speak about a theory of agronomy means to speak about the discovered and understood mechanisms of life and the development of plants, animals, and microörganisms.

Essentially important to our agricultural science during the last half-century is a methodological standard of biological knowledge, the position of biological sciences on the principles of life and development in plant and animal forms, i.e., primarily the science called genetics.

(2) THE HISTORY OF BIOLOGY—ARENA OF IDEOLOGICAL CONFLICT

The appearance of Darwin's doctrine stated in his book *Origin of Species* laid the basis for scientific biology.

The leading idea of Darwinian theory is the doctrine of natural and artificial selection. By means of a selection of variations useful to an organism there was produced and is being produced that adaptation which we observe in animate nature: in the structure of organisms and in their adaptability to the conditions of life. Through his theory of selection Darwin gave a rational explanation of purposefulness in nature. His concept of selection is scientific and true. In substance the doctrine of selection, in its most general form, is taken from the centuries-old practices of agriculturists and animal-breeders who, long before Darwin, were creating varieties of plants and breeds of animals by empirical means.

In his scientifically correct doctrine of selection, Darwin, through the prism of practice, examined and analyzed numerous facts accumulated by naturalists from nature itself. Agricultural practice served Darwin as the material basis on which he developed his theory of evolution, explaining the natural causes for the adaptation of organization in the organic world. This was a major conquest by mankind in its knowledge of living nature.

According to the evaluation by F. Engels, the knowledge of the interrelation of the processes being perfected in nature moved forward with gigantic strides, particularly as a result of three great discoveries: first, the discovery of the cell; second, the discovery of the transformation of energy; third, principally, the related evidence presented by Darwin that the organisms surrounding us at present, not excluding man, appeared as a result of a long process of development out of some few one-celled germs, and these germs, in their turn, were formed from a protoplasm or albumen arising by chemical means.[1]

Highly regarding the significance of Darwin's theory, Marxist classicists at the same time indicated the errors allowed by Darwin. Darwin's theory, appearing indisputably materialistic in its basic features, contains within itself a series of substantial errors. Thus, for instance, it was a great blunder for Darwin to introduce reactionary Malthusian ideas into his theory of evolution, side by side with its materialistic principle. In our time this great blunder is magnified by reactionary biologists.

Darwin himself indicated his acceptance of the Malthus scheme. He writes about it in his autobiography:

> In October 1838, that is, fifteen months after I had begun my systematic inquiry, I happened to read for amusement Malthus on population, and, being well prepared to appreciate the struggle for existence which everywhere goes on from long-continued observation of the habits of animals

[1] F. Engels, "Ludwig Feierbach" in K. Marx and F. Engels, *Works*, XIV (1931), 665-66.

and plants, it at once struck me that under these circumstances favorable variations would tend to be preserved, and unfavorable ones to be destroyed. The result of this would be the formation of new species. Here then *I had at last got a theory by which to work.*[2] (My emphasis—T. L.)

Darwin's error of transferring into his own doctrine the mad-brained reactionary Malthusian scheme on populations is up to the present not realized by many. A true scientist-biologist cannot and must not ignore the erroneous parts of Darwin's doctrine.

Biologists must again and again consider carefully the words of Engels:

> All of Darwin's teaching about the struggle for existence is quite plainly a transference from society into the field of living nature of Hobbes's doctrine of the war of all against all and the bourgeois-economic doctrine of competition, side by side with Malthus' theory of populations. Having brought this to a focus (whose unconditional accuracy I dispute, as was already indicated in the first point, particularly in connection with Malthus' theory), the very same theories are transferred from organic nature back again to history and then are maintained as if they were proven to have the force of eternal laws of human society. The naïveté of this procedure *strikes one between the eyes,* it is not worth while to waste words on it. But if I did want to consider it in detail then I would do it thus, that first of all I would show that they are poor *economists,* and only then already poor naturalists and philosophers.[3]

For purposes of the propaganda of his reactionary ideas Malthus devised, so to speak, a natural law. "The cause to which I allude," writes Malthus, "is the constant tendency in all animated life to increase beyond the nourishment prepared for it."[4]

It must be clear to a progressively thinking Darwinist that, although the reactionary Malthusian scheme was adopted by Darwin, nevertheless essentially it contradicts the materialistic principles of his own doctrine. It is easy to notice that Darwin himself, being a great naturalist who had laid the foundation of scientific biology and created an epoch in science, could not be satisfied with Malthus' scheme which he accepted, but which as a matter of fact essentially contradicts the phenomena of living nature.[5]

Thus, under the pressure of an enormous number of biological

[2] C. Darwin, *Works,* I (Lepkovski's edition; 1907), 31.

[3] K. Marx and F. Engels, *Works,* XXVI (1935), 406-9. Engels' letter to P. L. Lavrov.

[4] T. Malthus, *Test of the Law of Populations,* p. 31. 1908.

[5] It should be noted that Lysenko only denounces Malthus. At no place does he attempt to show *how* Malthus' scheme contradicts Darwinism. Malthus is refuted apparently by derogatory epithets alone.

facts which he collected, Darwin was in several instances forced to modify basically the concept "struggle for existence" and to enlarge it considerably, even declaring it to be a metaphoric expression.

In his time Darwin was unable to free himself from the theoretical mistakes which he committed. These errors were discovered and pointed out by Marxist classicists. At present it is quite intolerable to accept the erroneous parts of Darwinian theory, based on the Malthusian scheme of populations with, so to speak, the resultant intraspecies struggle. So much the more intolerable is it to project the erroneous parts of Darwin's doctrine as the foundation stone of Darwinism (I. I. Schmalhausen, B. M. Zavadovski, P. M. Zhukovski). Such an approach to Darwin's theory hinders the creative development of the scientific nucleus of Darwinism.[6]

At the very outset of the appearance of Darwin's doctrine it became immediately apparent that the scientific, materialistic nucleus of Darwinism—the doctrine of the development of living nature—is in antagonistic contradiction with the idealism predominant in biology.

Progressive, thinking biologists, both ours and foreign, saw in Darwinism the only correct course for the further development of scientific biology. They undertook the active defense of Darwinism against the attacks of the reactionaries, especially against the church and against Bateson's scientific obscurantism.

Such eminent Darwinian biologists as V. O. Kovalevski, I. I. Mechnikov, I. M. Sechenov, and particularly K. A. Timiryazev defended and developed Darwinism with all the inherent passion of true scientists.

K. A. Timiryazev, as a great investigator-biologist, saw distinctly that a successful development of the knowledge of the life of plants and animals is possible only on the foundations of Darwinism, and that only on a basis of a Darwinism developed and raised to a new height can biology acquire the opportunity to help the farmer produce two stalks where but one is growing today.

If Darwinism in the form in which it appeared from Darwin's pen found itself in contradiction to the idealistic world outlook, then the development of the materialistic doctrine deepened this contradiction even more. Therefore, reactionary biologists did everything within their power to throw out of Darwinism its materialistic elements. The individual voices of progressive biologists like K. A. Timiryazev were submerged by the united chorus

[6] Lysenko again omits the "how."

of the anti-Darwinists in the camp of the reactionary biologists throughout the world.[7]

In the post-Darwinian period the overwhelming part of the world's biologists, instead of developing further Darwin's doctrine, did everything to vulgarize Darwinism and to destroy its scientific foundation. The most vivid evidence of this vulgarization of Darwin is in the doctrines of Weismann, Mendel, and Morgan, founders of modern reactionary genetics.

(3) TWO WORLDS—TWO IDEOLOGIES IN BIOLOGY

Having emerged at the transition eras—the Past and the Present—Weismannism and its subsequent Mendelism-Morganism directed their "sword" against the materialistic bases of Darwin's theory of development.

Weismann called his conception neo-Darwinism, but essentially it was a complete negation of the materialistic aspects of Darwinism and surreptitiously introduced idealism and metaphysics into biology.[8]

The materialistic theory of the development of living nature is impossible without the recognition of the necesisty of the heredity of individual differences acquired by an organism under certain conditions in its life, impossible without the recognition of the inheritance of acquired characteristics. Yet Weismann undertook to refute this materialistic condition. In his basic work, *Lectures on the Theory of Evolution,* Weismann declares "that such a form of heredity is not only unproven but that it is impossible even theoretically. . . ."[9] Referring to his own much earlier statements of a like nature, Weismann states that "by this, war was declared against Lamarck's principle, the directly altering influence of usage and non-usage, in fact this began the war which continues even down to our time, the war between the neo-Lamarckists and the neo-Darwinists, as the contending parties have been named."[10]

As we see, Weismann speaks of his declaration of war against Lamarck's principle, but it is not difficult to see that he declared war against that without which there is no materialistic theory of evolution, declared war against the materialistic foundations of Darwinism under a screen of words on "neo-Darwinism."

Rejecting the inheritance of acquired characteristics, Weis-

[7] The two preceding statements are false. For the Russian usage of such words as "materialism" and "Darwinism," see page 9.

[8] See page 10.

[9] A. Weismann, *Lectures on the Theory of Evolution*, p. 294. 1905.

[10] *Ibid.*, p. 294.

mann invented a special hereditary substance, stating that it is expedient "to seek the hereditary substance in the nucleus"[11] and that *"the sought-for bearer of heredity is included within the substance of the chromosomes"*[12] containing the rudiments, each of which "determines a definite part of the organism in its appearance and in its final form."[13]

Weismann asserts that "there are two large categories of living matter: *hereditary,* or *idioplasm; and "nutrient," or "trophoplasm. . . ."*[14] Further, Weismann declares that the bearers of the hereditary substance *"the chromosomes, represent so to speak a separate world,"*[15] autonomous from the body of the organism and its conditions of life.

Having reduced the living body to only a feeding base for the hereditary substance, Weismann then proclaims the hereditary substance to be immortal and never arising *de novo.*

He asserts:

> The germ plasm of a species is thus never formed *de novo,* but it grows and increases ceaselessly; it is handed down from one generation to another. . . . If these conditions be considered from the point of view of reproduction, the germ cells appear the most important part of the individual, for they alone maintain the species, and the body sinks down almost to the level of a mere cradle for the germ cells, a place in which they are formed, and under favorable conditions are nourished, multiply, and attain to maturity.[16]

The living body and its cells, according to Weismann, are but the container and nutritive medium of the hereditary substance; they themselves can never produce the latter, they "can never bring forth germ cells."[17]

Thus, a mythical hereditary substance is endowed by Weismann with the characteristic of continuous existence, not experiencing development, yet at the same time controlling the development of a perishable body.

"Further . . . the hereditary substance of the germ cell prior to the reduction division potentially contains all the elements of the body."[18] And although Weismann does state that "the germ plasm no more contains the determinants of a 'crooked nose' than it does those of a butterfly's tailed wing," he goes on to empha-

[11] *Ibid.,* p. 410.
[12] *Ibid.,* p. 411.
[13] *Ibid.,* p. 452.
[14] *Ibid.,* p. 413.
[15] *Ibid.,* p. 353.
[16] *Ibid.,* p. 505.
[17] *Ibid.,* p. 504.
[18] *Ibid.,* p. 419.

size that, nevertheless, germ plasm "contains a number of determinants which so control the whole cell-group in all its successive stages, leading on to the development of the butterfly's wing with all its veins, membranes, tracheae, glandular cells, scales, pigment deposits, and pointed tail, arises through the successive interposition of numerous determinants in the course of cell multiplication."[19]

Thus, according to Weismann, the hereditary substance does not know regeneration, and in the development of the individual the hereditary substance does *not* know development, *cannot* endure any dependent variations.

The immortal hereditary substance, independent of qualitative characteristics of development of the living body, controlling the perishable body yet not produceable by it—such essentially is the undisguised idealistic, mystical concept of Weismann, advanced by him under the guise of words on "neo-Darwinism."

Mendelism-Morganism completely absorbed and, it can be said, even magnified this mystical Weismannist scheme.[20]

Turning to the study of heredity, Morgan, Johannsen, and other pillars of Mendel-Morganism declared from the outset that they intended to investigate the phenomena of heredity independently of Darwin's theory of development. Johannsen, for instance, wrote in his fundamental work: ". . . one of the important tasks of our work was to put an end to the detrimental dependence of theories of heredity on speculations in the field of evolution."[21] The Morganists made such declarations in order to conclude their investigations with statements implying in the final analysis a denial of development in living nature or recognition of development as a process of purely quantitative variations.

As we noted earlier, the clash of the materialistic and idealistic world outlooks in biological science has taken place throughout all of its history.

Now, in an era of conflict between two worlds, two antithetical, mutually contradictory trends which affect almost all of the biological sciences have become especially sharply defined.

Socialistic agriculture and the collective state farm system have in principle created a new, their own, Michurinist, Soviet, bio-

[19] *Ibid.*, p. 466.

[20] Weismann's concept of the continuity of the germ plasm is accepted today. The "immortality" of the germ plasm means only that living cells are derived from living cells and divide in turn to produce more living cells. There is no break in the chain until a species becomes extinct. Weismann's hypothesis of the mechanism of cellular differentiation, however, is obsolete, and geneticists do *not* accept it. Lysenko's statement to the contrary is false.

[21] Johannsen, *Elements of Exact Investigation of Variability and Heredity*, p. 178. 1933.

logical science which is developing in close harmony with agronomic practice as an agronomic biology.

The foundations of Soviet agrobiological science were laid by Michurin and Williams. They generalized and developed all the best that was accumulated by science and practical work in the past. Through their works they added in principle much that is new to the knowledge of the nature of plants and soil, to the knowledge of agriculture.

The close association of science with collective state farm practices produces inexhaustible possibilities for the development of theory, for an ever-growing knowledge of the nature of living organisms and of the soil.

It is not an exaggeration to state that the feeble metaphysical Morganist "science" of the nature of living organisms can in no way be compared to our efficient Michurinist agrobiological science.

The new effective trend in biology, rather the new Soviet biology, agrobiology, is impaled on bayonets by the representatives of reactionary alien biology, and even by a group of scientists in our country.

The representatives of reactionary biological science, called neo-Darwinists, Weismannists, or, which is really the same, Mendelists-Morganists, defend the so-called chromosome theory of heredity.

The Mendelists-Morganists, in Weismann's footsteps, assert that there exists a certain special "hereditary substance" in the chromosomes, residing in the body of an organism, as if in a box and is transmitted to subsequent generations independent of the qualitative specificity of the body and of its conditions of life. It follows from this concept that the new tendencies and differences acquired by an organism under certain conditions of its development and life cannot be hereditary, cannot have evolutionary significance.

According to this theory, properties acquired by plant and animal organisms cannot be transmitted to offspring, *cannot be inherited.*

Mendelist-Morganist theory does not include the conditions of the life of a body in the content of the scientific concept "living body." The environment, according to the Morganists, is but a background, although a necessary one for the manifestation of these and other characteristics of a living body according to its heredity. Thus, from their point of view the qualitative variations in the heredity (nature) of living bodies are entirely independent of the conditions of the external environment, of the conditions of life.

The representatives of neo-Darwinism, the Mendelists-Morganists, regard as entirely unscientific the aspiration of investigators to control the heredity of organisms by means of a corresponding alteration of the conditions of life of these organisms. That is why the Mendelists-Morganists call the Michurinist orientation in agrobiology neo-Lamarckism, which in their judgment is entirely faulty and unscientific.

In reality, the situation is just the reverse.

In the first place, the well-known propositions of Lamarckism, which credit the active role of the external environmental conditions in forming the living body and the heredity of acquired characteristics, are not in the least faulty but, on the contrary, are entirely correct and thoroughly scientific, in contrast with the metaphysics of neo-Darwinism (Weismannism).[22]

In the second place it is not at all possible to call the Michurin concept either neo-Lamarckism or neo-Darwinism. It appears as a creative Soviet Darwinism, refuting the errors of both, and is free from the faults of that part of Darwin's theory which is concerned with Malthus' erroneous scheme.

It is impossible to deny that in the dispute, bursting into flame at the beginning of the twentieth century, between he Weismannists and the Lamarckists, the latter were closer to the truth because they defended the interests of science, whereas the Weismannists became dazzled by mysticism and broke with science.[23]

The true ideological aspects of Morgan genetics were (unknown to our Morganists) excellently revealed by the physicist E. Schrödinger.[24] In his book *What is Life? Physical Aspects of the Living Cell,* setting forth approvingly Weismann's chromosome theory, he arrived at a series of philosophic deductions. Here is an important one: ". . . the personal self equals the omnipresent, all-comprehending eternal self." Schrödinger regards this conclusion as "the closest the biologist can get to proving God and immortality at one stroke."[25]

[22] A characteristic *ex cathedra* pronouncement.

[23] See note 22.

[24] Lysenko here illustrates Aesop's fable of the wolf and the lamb. Even if a physicist said it, the geneticists are to be blamed for it. Erwin Schrödinger, a distinguished physicist, apparently yielded to an impulse to write on a biological subject, and like a number of other distinguished physicists in like circumstances, he was tempted to see ghosts in material with which he was not too familiar. In his description of the cell he included an account of mitosis, meiosis, and elementary genetics such as he could have got from any good textbook. This apparently made him a Weismannist-Mendelist-Morganist, and, of course, Lysenko maintains that the geneticists are to be held responsible for any opinions he expresses.

[25] E. Schrödinger, *What is Life? Physical Aspects of the Living Cell,* p. 123. Moscow, Publishing House for Foreign Literature, 1947.

We, the representatives of the Soviet Michurinist concept, maintain that the inheritance of characteristics acquired by plants and animals in the process of their development is possible and necessary. Ivan Vladimirovich Michurin mastered these possibilities on the basis of his experimental and practical work. Most important is the fact that Michurin's doctrine, stated in his works, opens to every biologist the means for controlling the nature of plant and animal organisms, the means for modifying it in a direction necessary for practical work through a control of the conditions of life, i.e., through physiology.

The sharply quickened conflict, having divided biologists into two irreconcilable camps, thus burst into flame over the old question: *Is the inheritance of characteristics and properties acquired by plant and animal organisms in the course of their lives possible?* In other words, does the qualitative variation of the nature of plant and animal organisms depend on the nature of the conditions of life influencing the living body, the organism?

The Michurin doctrine, essentially materialistic-dialectic, maintains from evidence that there is such a relation.

The Mendelist-Morganist doctrine, essentially metaphysical-idealistic, without evidence, denies such a relation.

(4) SCHOLASTICISM OF MENDELIAN-MORGANISM

At the foundation of the chromosome theory lies Weismann's absurd thesis of the continuity of the germ plasm and its independence from the soma, which was long ago condemned by K. A. Timiryazev. The Morganists-Mendelists, following Weismann, proceed from the point that genetic parents are not the parents of their offspring. Parents and offspring, in accordance with their doctrine, are brothers or sisters.

Moreover, both the former (i.e., the parents) and the latter, (i.e., the offspring) in general do not appear themselves. They are only by-products of an inexhaustible and immortal germ plasm. In the sense of variability the offspring are entirely independent from the germ-plasm's by-product, i.e., from the body of the parents.

Let us turn to a source such as an encyclopedia where plainly the quintessence of the problem's nature is given.

T. H. Morgan, one of the founders of the chromosome theory, writes in the article entitled "Heredity":

> The germ-cells become later the essential parts of the ovary and testis respectively. In origin, therefore, *they are independent of the rest of the body and have never been a constituent part of it. . . . Evolution is ger-*

minal in origin and not somatic as had been earlier taught. (My emphasis. T. L.) This idea of the origin of new characters is held almost universally today by biologists.

Castle expresses the very same thing, only in another version, in the article "Genetics" in the same *Encyclopedia Americana*. Stating that usually the organism develops from a fertilized egg, Castle further sets forth the "scientific" bases of genetics. Let us cite them:

> In reality the parent does not produce the child nor even the reproductive cell which functions in its origin. The parent is himself merely a by-product of the fertilized egg (or zygote) out of which he arose. The direct product of the zygote is other reproductive cells similar to those from which it arose. . . . Hence heredity (that is, the resemblance between parent and child) depends upon the close connection between the reproductive cells which formed the parent and those which formed the child, one being the immediate and direct product of the other. This principle of the "continuity of the germinal substance" (reproductive cell material) is one of the foundation principles of genetics. It shows why body changes produced in a parent by environmental influences are not inherited by the offspring. It is because offspring are not the product of the parent's body but only of the germinal substance which that body harbors. . . . To August Weismann belongs the credit for first making this clear. He may thus be regarded as one of the founders of genetics.

It is quite clear to us that the fundamental concepts of Mendelism-Morganism are false. They do not reflect the reality of living nature and represent an example of metaphysics and idealism.

In consequence of this evidence the Mendelists-Morganists of the Soviet Union, although sharing entirely verbatim the principles of Mendelism-Morganism, often shyly conceal and veil them, and screen their metaphysics and idealism with a verbal jacket. They do this in fear of being ridiculed by Soviet readers and audiences who know well that the first rudiments of organisms, or sex cells, are one of the results of the life-activity of the parent.

The chromosome theory of heredity may seem systematic, and perhaps in some certain degree true, only if we ignore the fundamental ideas of Mendelism-Morganism, and people not acquainted in detail with the life and development of plants and animals may be deceived by it. To refute it, we have only to assume as an absolutely true and well-known fact that the sex cells or rudiments of new organisms are produced by the organism itself, by its body and not by the sex cells from which the mature, parental organism arose. This completely upsets at one stroke all of the "systematic" chromosome theory.[26]

It is self-explanatory that the biological role and significance of chromosomes in the development of cells and organism is, of course, not in the least denied by what has been said, but this is not at all the role which has been attributed to chromosomes by Morganists.

Many examples can be cited in support of the charge that our native Mendelists-Morganists share the chromosome theory of heredity, its Weismannist basis and idealistic conclusions in entirety.

Thus Academician N. K. Koltsov stated: "Chemically the genotype with its genes remains constant throughout the course of all oögenesis and is not subject to metabolism, to oxidizing and reducing processes."[27] In this statement, absolutely inadmissible to a literate biologist, metabolism is excluded in one of the living parts of developing cells. Is it not clear to all that N. K. Koltsov's deduction is in full agreement with Weismannist, Morganist, idealistic metaphysics?[28]

N. K. Koltsov's inaccurate statement dates from 1938. It was exposed long ago by Michurinists. Possibly it would not be necessary to turn to former years if the Morganists did not continue to maintain precisely those same antiscientific attitudes to this very day.

For better proof of what has been indicated, let us turn to the aforementioned book by Schrödinger. In it the author writes essentially the same thing as Koltsov. Schrödinger, sharing the idealistic concept of the Morganists, likewise declares that there is an *"hereditary substance largely withdrawn from the disorder of heat motion.* (My underscoring. T. L.)"[29]

Schrödinger's translator, A. A. Malinovski (scientific collaborator in N. P. Dubinin's laboratory), in his concluding remarks to the book joins in full accord with Haldane's opinion linking the idea stated by Schrödinger with N. K. Koltsov's views.

A. A. Malinovski writes in 1947 in the indicated postscript: "The view taken by Schrödinger that a chromosome is a giant molecule (Schrödinger's 'aphroditic crystal') was first introduced

[26] Perhaps no paragraph illustrates better than the preceding Lysenko's complete ignorance of scientific methods and his utter lack of intellectual standards. To refute an unwanted theory (the chromosome theory of heredity) he finds it only necessary "to assume" a competing hypothesis "as an absolutely true and well known fact." Q. E. D. The hypothesis he accepts is our ancient friend "pangenesis."

[27] N. K. Koltsov, "Structure of Chromosomes and their Metabolism," *Biological Journal*, 7:1:38, p. 42.

[28] Again the wolf and the lamb.

[29] *Op. cit.*, p. 119.

by the Soviet biologist Professor N. K. Koltsov, and not by Delbrück, with whose name Schrödinger associates this concept."[30]

In any case it is not worth while to analyze the question as to the priority of authorship of this scholasticism. More pertinent is the high evaluation of Schrödinger's book given by one of our home-grown Morganists, A. A. Malinovski.

I will quote several excerpts from that eulogistic appraisal:

> In his book Schrödinger in a form, absorbing and accessible both to the physicist and to the biologist, reveals to the reader a new, rapidly developing course in science, connecting in significant manner the methods of physics and biology. . . .[31]
>
> Schrödinger's book itself represents, strictly speaking, the first cohesive results of this course. . . . Schrödinger introduces into this new course of science on life a great personal contribution which in a notable degree justified the enthusiastic reviews which his book received in the foreign scientific press.[32]

Since I am not a physicist I will not attempt to speak on the methods of physics, linked by Schrödinger with biology. In regard to biology, however, Schrödinger's book is genuinely Morganist, and it is this, precisely, that brings about Malinovski's admiration.

The delights spouted by the author of the concluding remarks in reference to Schrödinger speak very eloquently on the idealistic views and positions in biology of our Morganists.

B. M. Zavadovski, professor of biology at Moscow University, writes in his article "The Creative Role of Thomas Hunt Morgan": "Weismann's ideas found a wide response among biologists, and many of them followed the course prompted by this richly talented investigator. . . . Thomas Hunt Morgan was among those who valued highly the fundamental content of Weismann's ideas."[33]

About what "fundamental content" is there talk here?

It has to do with a very important idea both from the viewpoint of Weismann and of all Mendelist-Morganists, among these being Professor B. M. Zavadovski. Professor Zavadovski formulates the idea thus: "Which came first, the hen's egg or the hen?" "And to this pointed statement of a problem," writes Professor Zavadovski, "Weismann gives a clear, categorical reply: 'The egg.' "[34]

[30] A. Malinovski, Translator's Postscript, *ibid.*, p. 193.
[31] *Ibid.*, p. 130.
[32] *Ibid.*, p. 131.
[33] *Bulletin of the Moscow Society of Naturalists*, 52:3, 1947, p. 86.
[34] *Bulletin of the Moscow Society of Naturalists*, 52:4, 1947, p. 83.

It is clear that both the question and the reply given by Professor Zavadovski's repetition of Weismann are but a simple and belated revival of ancient scholasticism.

In 1947 Professor B. M. Zavadovski repeats and defends the same ideas which he expressed in 1931 in his work *The Dynamics of Development of an Organism*. B. M. Zavadovski considered it necessary "to add firmly his voice to that of *Nussbaum* who states that sex products are developed not from a maternal organism but from an identical source from which it came,"[35] that "spermatozoa and ova do not originate in the parental organism but have a common origin with the latter."[36] And in the "general conclusions" of his work Professor Zavadovski wrote: "Analysis leads us to the conclusion that it is impossible to consider the cells of the embryo as derivatives of somatic tissues. It is necessary to consider the germ cells and the somatic cells not as a daughter and parent generation but as twin-sisters of which one (the soma) appears as the nourisher, defender, and guardian of the other."[37]

Geneticist N. P. Dubinin, Professor of Biology, wrote in his article "Genetics and Neo-Darwinism": "Perfectly justified, genetics divides an organism into two separate divisions, the hereditary plasma and the soma. Moreover, this division is one of its fundamental conditions, one of its outstanding generalizations."[38]

We shall not prolong further the list of such outspoken authors as B. M. Zavadovski and N. P. Dubinin, who express the ABC's of the Morgan system of opinion. These ABC's are known in university textbooks of genetics as the principles and laws of Mendelism (law of dominance, principle of segregation, principle of pure gametes, etc.). An example of how far our native Mendelists-Morganists uncritically accept idealistic genetics is the fact that until recently the basic textbook on genetics in many of our higher institutes of learning is a translation of the strictly Morganist American textbook of Sinnott and Dunn.

In conformity with the fundamental positions of this textbook Professor N. P. Dubinin in his article "Genetics and Neo-Lamarckism" wrote: "Thus the facts of modern genetics can in no degree be reconciled with a recognition of the 'principle of principles' of Lamarckism, *with the idea of the inheritance of acquired characteristics*.[39] (My underscoring. T. L.)"

Thus, the circumstance of the possibility of the inheritance of acquired deviations, one of the greatest acquisitions in the history

[35] B. M. Zavadovski, *The Dynamics of Development of an Organism*, p. 321. 1931.
[36] *Ibid.*, p. 313.
[37] *Ibid.*, p. 326.
[38] *Journal of Natural Science and Marxism*, 1929, No. 4, p. 83.
[39] *Ibid.*, p. 81.

of biological science, whose foundation was laid by Lamarck and was organically appropriated later in Darwin's doctrine, was thrown overboard by the Mendelists-Morganists.

And so, against the materialistic doctrine of the possibility of inheritance by plants and animals of individual deviations in characteristics acquired under certain conditions of life, Mendelism-Morganism made the idealistic assertion that the living body, the so-called soma, is separated from an immortal hereditary substance, the germ plasm. In addition, it is categorically stated that an alteration of the soma, the living body, has no influence whatsoever on the hereditary substance.

(5) THE IDEA OF UNKNOWABILITY IN THE CONCEPT OF "HEREDITARY SUBSTANCE"

Mendelism-Morganism imparts the postulated mythical "hereditary substance" with a vague character of variability. Mutations, i.e., variations of the "hereditary substance," allegedly do not have a definite direction. This statement of the Morganists is logically associated with the basic principle of Mendelism-Morganism, with the circumstance of the independence of the hereditary substance from the living body and its conditions of life.

Proclaiming the "indefiniteness" of hereditary variations, the so-called "mutations," Morganists-Mendelists conceive of hereditary variations as *fundamentally unpredictable.* This is a unique conception of unknowability and its name is idealism in biology.

The assertion about "indefiniteness" of variability closes the door to scientific foresight and with it impairs agricultural practice.[40]

Proceeding from the unscientific, reactionary doctrine of Morganism on "indefinite variability," the Director of the Department of Darwinism at Moscow University, Academician I. I. Schmalhausen, states in his work *Factors of Evolution* that, in its specificity, hereditary variability is not dependent on the conditions of life and therefore is deprived of direction.

". . . The factors not utilized by an organism," writes Schmalhausen, "if generally they reach the organism and influence it,

[40] This statement illustrates Lysenko's complete inability to understand the simplest principles of the mathematics of chance. Fortuitous variations were all that Darwin needed for the raw material from which nature could select. Modern geneticists likewise take advantage of chance mutations; they discard those which are undesirable and preserve those which are useful. For adaptative mutations (other than those which arise by chance) to occur in animals and plants in novel surroundings implies a *foresight* in evolution. It will be interesting to learn how this ancient bit of mysticism will be converted to "materialism."

can have only an indefinite effect. Such an effect can be only indeterminate. Consequently all new variations of the organism which as yet have no past history will be indeterminate. This category of variations, however, will include not only mutations, as new "hereditary" variations, but also any new modifications, i.e., those arising for the first time."[41]

On a preceding page Schmalhausen writes: "In the development of any individual, the factors of the external environment appear only in the role of agents liberating the course of well-known form-producing processes and conditions which make it possible to complete their realization."[42]

This formalistic, auto-mystical theory of the "liberating principle," wherein the role of external conditions is reduced only to the realization of an autonomous process, has long been smashed by the progressive movement of advanced science and has been exposed by materialism as unscientific and idealistic in its nature.

In addition Schmalhausen and other native followers of alien Morganism refer to Darwin in their statements on the subject. Proclaiming "indefiniteness of variability," they clutch after the corresponding expressions of C. Darwin on this question. Actually, Darwin did speak about "indefinite variability." But Darwin's declarations had in their basis the *limitations* of the selective practice of his time. Darwin himself accounted for this and wrote: "We cannot at present explain either the causes or the nature of variability in organic beings.[43] "This problem is obscure, but it is possible; we have to be convinced of our ignorance."[44]

The Mendelists-Morganists grasp all that is obsolete and inaccurate in Darwin's doctrine, at the same time scorning its vital materialistic nucleus.

In our socialistic land the teaching of the great transformer of nature I. V. Michurin created on principle a new foundation for the control of the variability of living organisms.

Michurin himself and his followers, Michurinists, have produced and are producing controlled hereditary variations of plant organisms literally in mass quantity. Notwithstanding all this, Schmalhausen even now states on the given question:

> The appearance of individual mutations has all the characteristics of chance phenomena. We cannot foretell or bring about arbitrarily one or another mutation. So far all attempts to establish any regular connection

[41] *Factors of Evolution,* pp. 12-13. Pub. Acad. of Sci., USSR, 1946.

[42] *Ibid.,* p. 11.

[43] C. Darwin, *Variation of Animals and Plants in Domestic Condition,* p. 479. State Agricultural Pub. House, 1941.

[44] *Ibid.,* p. 452.

between the quality of a mutation and a specific variation in the factors of the external environment has failed.[45]

Proceeding from the Morganist conception of mutations, Schmalhausen has declared a theory of so-called "stabilizing selection," which is profoundly untrue ideologically and detrimental in practical work. According to Schmalhausen, breed-formation and variety-formation proceed inevitably in accordance with a disappearing curve: turbulent at the dawn of culture, breed and variety formation evermore squanders its "reserve of mutations" and gradually slows down. "Both breed formation of domestic animals and variety formation of cultivated plants," writes Schmalhausen, "proceeded with exceptional rapidity, obviously at the expense of an earlier accumulated reserve of variability. Further rigidly directed selection is already proceeding more slowly."[46]

Schmalhausen's statements and his entire conception of "stabilizing selection" are pro-Morganist.[47]

As is well known, Michurin created in the course of a single human life more than three hundred plant varieties. An entire group of them was produced without sexual hybridization, and all were produced by means of a rigidly applied selection including systematic training. In view of these facts and the continued achievements of the followers of Michurin's doctrine, the maintenance of a slowing down of rigidly directed selection is futile fabrication against advanced science.[48]

Michurinist facts, it seems, confuse Schmalhausen in the exposition of his theory of "stabilizing selection." In his book *Factors of Evolution,* he escapes the difficulty by passing over completely any mention of Michurinist works and even of the very existence of Michurin as a scientist. Schmalhausen wrote a thick book on the factors of evolution, yet nowhere and not once, not even in the bibliography, did he mention either K. A. Timiryazev or I. V. Michurin. And yet K. A. Timiryazev left to Soviet science a remarkable theoretical work which directly bears the title

[45] *Factors of Evolution,* p. 68. Pub. Acad. of Sci., USSR, 1946.

[46] *Ibid.,* pp. 214-15.

[47] A little explanation is indicated here. Any species in nature accumulates a store of mutations, particularly mutations which have a neutral survival value or which are passed on as recessives. The artificial selection to which domesticated animals and plants are subjected uses up this accumulation. New improvements wait upon new mutations. The practical advantages of widespread hybridization lie in the possibility of combining advantageous mutations which have occurred in different stocks and have been selected in different regions.

[48] Michurin was an importer of plants. Some of his reported hybrids were not hybrids at all (see page 71). Until Michurin's data are published in a form that allows them to be checked, geneticists will remain skeptical as to his claims.

Factors of Organic Evolution, while Michurin and Michurinists have placed the factors of evolution at the service of agricultural economy, revealing new factors and strengthening the understanding of old ones.

"Having forgotten" about foremost Soviet scientists, about the founders of Soviet biological science, Schmalhausen at the very same time, frequently and with emphasis, alludes to and cites statements of important and minor foreign and native agents of Morganist metaphysics, the leaders of reactionary biology.

Such is the style of "Darwinist" Academician Schmalhausen. And this book was recommended at a meeting of the Faculty of Biology of Moscow University as a masterpiece of creative development of Darwinism. This book was highly valued by the deans of two biology faculties, of Moscow and Leningrad Universities; this book was praised by the Professor of Darwinism at Kharkov University, I. Polyakov; by the Pro-Rector of Leningrad University, J. Polyanski; by Academician B. M. Zavadovski of our Academy, and by a series of other Morganists who regard themselves as orthodox Darwinists.

(6) THE STERILITY OF MORGANISM-MENDELISM

Repeatedly and without proof, and often even slanderously, the Morganists-Weismannists, i.e., the supporters of the chromosome theory of heredity, have stated that I, as President of the Academy of Agricultural Sciences, have in the interests of the Michurinist movement in science supported by me suppressed administratively the other movement, opposed to the Michurinist.[49]

Unfortunately, until now the situation has been just the reverse, and in this respect it is possible and proper to blame me as president of the All-Union Academy of Agricultural Sciences. I was unable to find within myself the strength and understanding in sufficient measure to make use of the official position given me in the matter of creating the conditions toward a greater development of the Michurinist movement in the different branches of biological science, or to circumscribe even a little the scholasticists and metaphysicists of the opposing movement. Therefore, in reality it has turned out that until now the movement which was represented by the president, the Michurinist movement, has been suppressed specifically by the Morganists.

We Michurinists should frankly admit that until the present we have as yet not adequately succeeded in making use of all the

[49] It seems queer that, in the light of what follows, this accusation was considered slanderous.

excellent opportunities established in our country by the Party and the Government for a complete exposure of the Morganist metaphysics, which was introduced entirely from an inimical, alien, reactionary biology. The Academy, just recently supplemented by a significant number of Academician-Michurinists, is now obligated to fulfill this most important task.[50] This will be important scientifically in the task of cadre preparation, and in the task of increasing aid to the collective and state farms.

Morganism-Mendelism (the chromosome theory of heredity) is still taught in different variations in all the biological and agronomical institutes, while the teaching of Michurinist genetics has essentially, to all intents and purposes, not been introduced. Often even in the highest official scientific circles of biologists the adherents of the doctrines of Michurin and Williams were in the minority. Until the present time they have even been in the minority in the membership of the V. I. Lenin All-Union Academy of Agricultural Sciences. Thanks to the care of the Party, the Government, and personally of Comrade Stalin, the situation in the Academy has now been radically changed. Our Academy was enlarged and soon in the next elections it will be even more enlarged by a notable number of new academicians and corresponding-member Michurinists. This will establish a new arrangement and new possibilities for the future development of Michurinist teaching in the Academy.

The statement that the chromosome theory of heredity, at the basis of which lie true metaphysics and idealism, has been suppressed until now is absolutely false. Up to the present the situation has been precisely the opposite.

In our country the Michurinist movement in agrobiological science has stood and stands in its practical effectiveness athwart the path of the cytogeneticists-Morganists.

Aware of the practical worthlessness of the theoretical premises of their metaphysical "science," and not wishing to renounce them and accept the effective Michurin direction, the Morganists have, at the root of their antagonistic false science, applied and do apply all their efforts to the purpose of stopping the development of the Michurin movement.

The assertion that someone is preventing the cytogenetic direction in biological science from being applied to practical work in our country resounds with slander. Doubly wrong is he who speaks as if "the right to the practical application of the fruits of one's

[50] Here is a frank confession of packing the Academy with Michurinists. The reader should have little trouble in picking out the papers of these nouveau academicians which are published in the following chapters. Their naïveté and ignorance are really striking.

work were a monopoly of Academician Lysenko and his followers."

The Ministry of Agriculture could show just exactly what cytogeneticists have offered for application to practical work, and if such proposals were actually made whether they were accepted or rejected.

The Ministry of Agriculture could likewise show which of its scientific research institutes (not to mention the educational institutes) were not devoted to cytogenetics in general and polyploidy of plants, produced by means of the use of colchicine, in particular.

It is well known to me that many institutes have been devoted to and are occupied with this, in my opinion, wasteful work. In addition, the Ministry of Agriculture has disclosed a special institute for investigations on the problems of polyploidy, with A. R. Zhebrak at its head. I believe that this institute, occupying itself over a period of years only with this work, polyploidy, has literally produced nothing of value.

The worthlessness of the practical and theoretical endeavors of our native cytogeneticists, Morganists, can be illustrated simply in the following example.

One, and in the opinion of our Morganists, allegedly one of the most prolific of them, Professor-Geneticist N. P. Dubinin, member-correspondent of the Academy of Sciences, USSR, has for many years worked on the clarification of the differences in the cell nuclei of the fruit flies in the city and its rural districts.

For the sake of complete clearness, let us indicate the following: Dubinin does not study the qualitative changes, in the given instance of the cell nucleus, in their relation to the influence of different conditions of life according to quality. He does not study the inheritance of acquired differences by fruit flies under the influence of specific conditions of life, but the changes identifiable by chromosomes in the composition of a population of these flies, in consequence of a simple destruction of them, particularly during a period of war. Such destruction is called "selection" by Dubinin, as well as other Morganists. (Laughter) This sort of "selection," identical with a simple sieve and having nothing in common with its real creative role, appears as the theme of Dubinin's study.[51]

[51] Here Lysenko quotes passages from Dubinin out of context and leaves out those parts which tie the whole together. As quoted, of course it makes no sense. In brief, Dubinin was investigating the changes in the hereditary mechanism which accompany the formation of new species. These changes are generally greater than point mutations or gene replacements and usually involve such chromosome alterations as reversions, deletions, reduplications, etc. It happens that these changes can be investigated best in the giant salivary gland chromosomes of the *Diptera* (flies). Dubinin was using the best research material for the problem he was investigating.

This work is called *Structural Variability of Chromosomes in Populations of City and Rural Districts.*

I quote several excerpts from this work:

During examination of segregated populations of *D. funebris* the fact was noted in a work of 1937 of observable differences according to the concentration of inversions. Tinyakov emphasized this phenomenon on the basis of voluminous material. However, only the analysis of 1944-45 revealed to us that these intrinsic differences of populations are related to the differences in the conditions of habitation in city and country.

The population of Moscow has eight different arrangements of genes. In the second chromosome there are 4 arrangements (the standard and 3 different inversions). There is one inversion in chromosome III and one in IV. . . . Inversion II-1 has limits from 23C to 31B. Inversion II-2 from 29A to 32B. Inversion II-3 from 32B to 34C. Inversion III-1 from 50A to 56A. Inversion IV-1 from 67C to 73A.b. During the years 1943-45 the karyotype of 3315 individuals in the population of Moscow was studied. The population contained great concentrations of inversions which were found to be different in the various regions of Moscow.[52]

During and after the war Dubinin pursued his investigations, occupying himself with the problem of fruit flies in the city of Voronezh and its suburbs. He writes:

The destruction of the industrial centers in the course of the war disrupted normal living conditions. The populations of *Drosophila* found themselves under severe conditions of existence, which possibly exceeded the severity of hibernating places in rural regions. Unusually interesting was the study of the influence of changes in the conditions of existence produced by war on the karyotypic structure of the city's populations. During the spring of 1945 we studied the populations of Voronezh, one of those cities which suffered most severe destruction from German invasion. Among 225 individuals, only two flies were found heterozygotic, according to inversion II-2 (0.88%). Thus the concentrations of inversions in this large city were found to be lower than in certain rural districts. We see the catastrophic influence of natural selection on the karyotypic structure of a population.[53]

As we see, Dubinin reports his work so that externally it may appear as scientific to some. No wonder this work figured as one of the principal reasons in the election of Dubinin as corresponding member of the Academy of Sciences of the USSR.

But if we give a more simple account of this work, after freeing it of its wordy pseudoscientific mold and replacing the Morganist

[52] *Reports, Academy of Sciences, USSR,* 1946, 11:2, p. 152.
[53] *Ibid.,* p. 153.

jargon with ordinary Russian words, then the following will make itself apparent:

As a result of many years of work, Dubinin "enriched" science by the "discovery" that in a composition of a fly population among the fruit flies of Voronezh and its suburbs an increase in the percentage of flies with certain chromosomal differences and a decrease among other fruit flies with other differences in the chromosomes (in the Morgan jargon this is known as "concentration of inversion II-2") occurred during the war.

Dubinin does not limit himself to such "highly valuable" discoveries for theory and for practical work produced by himself during the time of the war; he is planning future undertakings even for the reconstruction period and writes:

"It will be very interesting to investigate in the course of a series of succeeding years the restoration of the karyotypic structure of a city population in relation to the restoration of normal conditions of life."[54] (*Stirring in the auditorium. Laughter*)

Such is a typical Morganist "contribution" to science and practical work up to and during the time of the war, and such the perspectives of Morganist "science" for the period of reconstruction! (*Applause*)

(7) MICHURIN'S DOCTRINE—FOUNDATION OF SCIENTIFIC BIOLOGY

In contrast to Mendelism-Morganism, which affirms that the causes of variability in the nature of organisms cannot be recognized and which denies the possibility of a directed alteration of the nature of plants and animals, I. V. Michurin's motto proclaims: "We cannot expect favors from Nature; our task is to take them from her."

On the basis of his work I. V. Michurin arrived at the following conclusion: "It is possible with human intervention *to bring about change* in every form of animal or plant *more quickly and in a direction desirable by man.* A broad field of utmost usefulness is opened to man. . . ."[55]

Michurin's teaching completely repudiates the fundamental position of Mendelism-Morganism, the position that there is complete separation of the hereditary characteristics of plants and animals from environmental influences. Michurin's teaching does not recognize the existence in an organism of an hereditary substance separate from the body of the organism. An alteration in the heredity of an organism or in the heredity of a separate

[54] *Ibid.*, p. 153.
[55] *Works*, IV, 72.

part of its body always appears as a result of a change of the living body itself.[56] A change in the living body takes place as a result of a deviation from the normal type of assimilation and dissimilation, and of a change or departure from the normal type of metabolism. This variation in the organisms or in their separate organs and properties is transmitted to the offspring, although not always or completely; nevertheless the modified rudiments of newly incipient organisms are never produced except as a result of a modification of the body of the parental organism, as a result of the direct or indirect influence of the environment on the development of an organism or of its separate parts, including the sexual and vegetative rudiments. The variation of heredity, the acquirement of new characteristics and their strengthening and accumulation in a series of successive generations, is always dependent on the environment of the organism. Heredity is modified and complicated by means of the accumulation of newly acquired characteristics and properties by organisms in a sequence of generations.

The organism and the conditions necessary for its life represent a unity. Different living bodies require different external environmental conditions for their development. Inquiring into the nature of these requirements, we likewise learn of the qualitative characteristics of heredity. Heredity is the *characteristic of the living body to require definite conditions for its existence, its development, and its definite response to the various conditions.*

A knowledge of the inherent requirements and the relationship of the organism to the conditions of the external environment presents the opportunity for controlling the existence and the development of this organism. Control of the conditions of life and develpment of plants and animals permits one to understand ever more deeply their nature, and through this to establish methods for its alteration in a direction needful to man. It is possible on the basis of a knowledge of the methods of control to alter directively the heredity of organisms.[57]

Every living body forms itself out of the conditions of the external environment for its own harmony, in accordance with its heredity. That is why different organisms live and develop in one and the same environment. As a rule each generation of plants or animals develops in the main like its predecessors, particularly the immediate ones. *Reproduction of forms similar to itself is a general characteristic feature of any living body.*

In those cases where the organism finds conditions in the sur-

[56] Again pangenesis! See note 26.

[57] Scientific evidence for this confession of faith is apparently not considered necessary.

rounding environment corresponding to its heredity, the development proceeds just as it had proceeded in preceding generations. When the organisms do not find all the conditions necessary for them and still must assimilate the conditions of an external environment, to some extent unsuited to their nature, there are produced whole organisms or separate parts of their bodies more or less different from the preceding generation. If the altered part of the body is original to the new generation, the latter will be altered in its needs and in its nature in some degree from the preceding generations.

The cause of a change in the nature of a living body is a change in the type of assimilation, of the type of metabolism. For instance, the process of vernalization of summer grain crops requires conditions of lowered temperatures for its accomplishment. The vernalization of summer grain crops proceeds normally at temperatures effective in the spring and summer under field conditions. But if the summer grain crops are vernalized under conditions of lowered temperatures, the vernalized plants can be transformed into winter grain in from two to three generations.[58] The winter grain crops cannot proceed through the processes of vernalization without the presence of lowered temperatures. This concrete example indicates by what means a new need is created in the offspring of given plants—the need of lowered temperatures for vernalization.

The sex and any other cells through which organisms reproduce are produced as a result of the development of the entire organism by means of transformation, by means of metabolism. The course of development through which an organism passes is, as it were, accumulated in the cells which produce a new generation.

Therefore it can be said: In whatever degree in the new generation there is built anew the body of this organism (let us say a plant) in that very degree are developed all of its characteristics, and among these heredity.

The development of different cells, of different types of cells, and the development of independent processes in one and the same organism require different conditions of the external environment.

In addition, these conditions are assimilated in various ways. It is necessary to emphasize that in a given case *that which is assimilated is understood to be external, while that which assimilates is internal.*

The life of an organism goes through a countless number of processes and transformations in conformity with the law of

[58] See page 27.

development. Food taken into the organism from the external environment is assimilated by the living body through a chain of different transformations; from the exterior it passes into the interior. This internal portion, being alive, exchanges substances with the other cells and parts of the body, nourishes them, and becomes in this manner external in regard to them.

Two kinds of qualitative changes are observed in the development of plant organisms:

1. Changes connected with the process of realization of an individual cycle of development when the intrinsic requirements, i.e., heredity, are normally satisfied *by the corresponding conditions of the external environment.* As a result a body of the very same kind and heredity is produced as in preceding generations.

2. Changes of kind, i.e., changes in heredity. These changes also appear as a result of individual development, but as deviating from the normal, usual course. Ordinarily change in heredity appears in consequence of the development of the organism *under conditions of the external environment in some measure not corresponding to the intrinsic needs of the organism.*

Changes of the conditions of life compel the very type of development of plant organisms to be changed. The species-change type of development appears thus as the original cause of a change of heredity. All of those organisms which cannot be changed in accordance with changed conditions of life do not survive, do not leave descendants.

Organisms, hence also their nature, are produced only through the process of development. Of course, the living body can also undergo change outside of development (burns, joint fracture, tearing off of roots, and so forth); these changes, however, will not be characteristic or necessary for the living process.

Numerous facts indicate that changes in different parts of the body of a plant or animal are frequently not passed on fully to the sex cells.

This is explained by the fact that the process of development of every organ, of every particle of a living body, requires relatively specific conditions of the external environment. These conditions are selected during development of each organ and of even the finest organelle from the environment surrounding them. Therefore, if this or that part of the body of a plant is forced to assimilate relatively unusual conditions which result in a change distinguishable from analogous parts of a preceding generation, then the substances passing from it to the neighboring cells may not be accepted or included by them in a further chain of characteristic processes. Association of the altered part of the body with other parts of the plant will, of course, take place; however, this association may not be completely reciprocal. The modified part of the body will continue to receive some sort of nutriment from the adjoining parts, yet it will not be able to donate its own specific substances to the whole as the neighboring parts will reject them.

Hence an often observable phenomenon is understandable, i.e., the instances when the changed organs, characteristics, or properties of an organism do not make their appearance in the descendants. But in themselves these changed parts of the body of the parental organisms always possess a changed heredity. The practice of orchard and flower cultivation has known these facts long ago. An altered twig or bud in a fruit tree or eye (bud) of a potato tuber, as a rule, cannot modify the heredity of the offspring of a tree or tuber which does not descend directly from it, from the altered parts of the parts of the parental organism. If, however, this modified part is grafted and raised as a separate, self-contained plant, it will possess, as a rule, the modified heredity, the heredity that was indigenous to the altered part of the parental body.[59] *The degree of hereditary transfer of variations will depend on the degree of inclusion of the substances of the altered part of a body into the common chain of the process leading toward the formation of reproducing sex or vegetative cells.*

Knowing the course of construction of the heredity of an organism, it is possible to vary it directively by means of the production of specific conditions at a specific moment of the development of an organism.

Excellent varieties of plants as well as excellent breeds of animals have always been produced and are being produced in practical work only under circumstances of an excellent agrotechnology and an excellent zoötechnology. With inferior agrotechnology it is not only never possible to produce good varieties from poor, but in many instances even excellent cultivated varieties become inferior in several generations under such circumstances. The fundamental rule in the practical work of seed growing says that crops on a seed section must be cultivated in as superior a manner as possible. To accomplish this it is necessary to establish, by means of agrotechnology, excellent conditions, suitable to the optimum of the hereditary requirements of given plants. The very best among well-cultivated plants must

[59] The two preceding paragraphs are really as bad as they sound. What Lysenko is trying to say is that acquired characters are not always inherited, but that bud mutations can be made to produce new strains through vegetative propagation. If he had used the terms "pangens" or "gemules," he could have saved himself much circumlocution.

be, and are, selected for seed. By this procedure plant varieties are perfected in practice. With inferior cultivation, however (i.e., with application of an inferior agrotechnology), no selection of superior plants for seed will give the necessary results. With such cultivation all the seeds become inferior, and the very best among the inferior will still be inferior.

The chromosome theory of heredity admits the possibility of producing hybrids only by sexual means. It rejects the possibility of producing vegetative hybrids, just as it does not recognize the specific influence of the conditions of life on the nature of plants. But I. V. Michurin not only recognized the possibility of the existence of vegetative hybrids, but even worked out the method of a mentor. The method consists of this, that by means of the grafting of cuttings (twigs) of any old variety of fruit tree to the crest of a young variety, characteristics not possessed by the young variety are acquired by it, are transmitted to it from the grafted twigs of the old variety. That is why the method indicated was called a mentor-trainer by I. V. Michurin. A wilding is likewise made use of in the role of mentor. By this means a series of new superior varieties was developed or improved by Michurin.

I. V. Michurin and the Michurinists found means for mass production of vegetative hybrids.

Vegetative hybrids are the conclusive proof of the correctness of the Michurinist conception of heredity. At the same time they represent an insurmountable obstacle to the theory of the Mendelists-Morganists.[60]

Organisms which have not been formed by stages, not having as yet gone through a complete cycle of development, will always modify their development at grafting in comparison with root-possessing, i.e., not grafted, plants. Through the union of plants by means of grafting, a single organism with a diverse nature is produced, namely, with the nature of both graft and stock. By collecting and sowing the seed from the scion or the stock it is possible to obtain offspring of plants, individual representatives of which will possess characteristics not only of the variety of which the seed had been taken, but also of the other with which it was united by means of grafting.[61]

It is obvious that the stock and the graft could not exchange chromosomes from the nuclei of cells, and that all hereditary characteristics were transmitted from the wilding to the graft

[60] Chimeras or graft hybrids have been known since the seventeenth century. They have been intensively studied by the early Mendelians and are at present treated in many textbooks of genetics and botany. They are curiosities without practical importance. Geneticists have not obtained the results claimed for Michurin.

[61] This does not occur west of the Iron Curtain.

and inversely. Therefore, *the plastic substances producible by scion and stock in the same way as chromosomes, or as any particle of a living body, possess breeding qualities, and a definite heredity is inherent in them.*

Any characteristic can be transmitted from one variety to another by means of grafting as well as by means of sex.

A large body of factual material on the vegetative transmission of various characteristics of potato, tomato, and a range of other plants leads to the conclusion that vegetative hybrids are in principle not distinguishable from sex hybrids.

The representatives of Mendel-Morgan genetics not only cannot produce controlled modifications of heredity, but they categorically deny the possibility of such a modification adapted to the influence of environmental conditions. Proceeding from the principles of Michurin's doctrine, it is possible to modify heredity *in complete conformity with the effects of the living conditions.*

As an instance of this we may cite the experiments on the transformation of spring grain into winter, and winter grain into an even more wintry type, for example, in Siberia where there are severe winters. These experiments have not only a theoretical but also an unusual practical interest in the production of winter-resisting varieties. There is already in existence a range of winter forms of wheat obtained from spring forms which cannot be surpassed in their resistance to frost, and several even surpass the most frost-resisting varieties known in practical work.

Numerous experiments show that in the liquidation of the old, established property of heredity, a settled, consolidated, new heredity is not secured at one stroke. In the vast majority of instances, organisms with a plastic nature, termed "unstable" by I. V. Michurin, are produced.

Plant organisms with an "unstable" nature are those in which their conservatism has been liquidated and their selectivity in relation to the conditions of the external environment weakened. In such plants, in place of a conservative heredity there is retained or appears anew only a tendency to select certain conditions before others.

It is possible to shake loose the nature of a plant organism:

(1) by means of grafting, i.e., union of the tissues of plants of different varieties;

(2) by means of the influence of conditions of the external environment at specific moments in the continuation of these or other processes in the development of an organism;

(3) by means of crossing, especially of forms sharply differentiated according to the place of their habitation or origin.

Great attention was paid by the best biologists to the practical significance of plant organisms with an unstable heredity in the first place and specifically by I. V. Michurin. The plastic plant types with unsettled heredity and produced by this or that method must be further cultivated in addition from generation to generation under those conditions to which it is necessary to adapt and to develop in order for us to secure a desired organism.

In the majority of plant and animal forms new generations are developed only following fertilization—the union of female and male sex cells. The biological significance of the process of fertilization consists of this, that in this manner organisms with a dual heredity are produced: maternal and paternal. This dual heredity produces the greater vitality of organisms and the much wider spread of their adaptiveness to the varying conditions of life.

The biological need for crossing forms, even if they differ slightly among themselves, is determined through the utility of the enrichment of heredity.

The rejuvenation and reinforcement of the vitality of plant forms can likewise proceed through vegetative, nonsexual means. It is achieved in a living body by means of the assimilation of new (unusual for it) conditions of the external environment. In an experimental setting—during vegetative hybridization, in tests for the production of spring from winter forms or winter from spring forms and in a series of other examples of upsetting the nature of organisms—it is possible to observe the rejuvenation and reinforcement of the vitality of organisms.

By controlling the conditions of the external environment, the conditions of life of plant organisms, it is possible directively to change and to produce varieties with a heredity needful to us.

Heredity is the effect of a concentration of the influences of conditions of the external environment, assimilated by organisms in a sequence of preceding generations.

It is possible through skillful hybridization and the union of varieties by sexual means to combine in one organism that which was assimilated and secured by many generations of varieties used for crossing. However, according to Michurin's doctrine not one hybridization will give positive results unless conditions suitable to the development of those characteristics which it is desired to produce in the developed or improved variety are created.

I have given an account of Michurin's doctrine only in general outline. Here it is important only to emphasize the absolute necessity for all Soviet biologists to study this doctrine more deeply. For scientific workers in the various branches of biology the best method for a mastery of the operative theoretical depths of Michurin's doctrine is the path of investigation, frequently repeated reading of Michurin's works, and a review of his individual works from the point of view of a practical solution of important problems.

Socialist agriculture is in need of a developed, profound biological theory which would quickly and accurately help to improve the agronomic methods of plant cultivation and the production of large, stable crops. It is in need of a profound biological theory which would aid agricultural workers to produce in the shortest time the desired highly productive forms of plants, responding in their nature to the high fertility established by collective-farm workers in their fields.

The unity of theory and practice is the true highway of Soviet science. Michurin's doctrine is exactly such a doctrine, embodying in a supreme form this unity in biological science.

I have frequently cited in my public appearances examples of the fruitful application of Michurin's doctrine in the practical solution of important problems in various branches of plant cultivation. In the present instance I will take the liberty to consider briefly only a few problems of animal husbandry.

Animals, like plants, were and are formed in intimate relationship with the conditions of their life, with the conditions of the external environment.

The basis for increasing the productivity of domestic animals, improving existing breeds, and developing new ones lies in food and the conditions of maintenance. This is particularly important for increasing the effectiveness of hybridization. Various breeds of domestic animals were developed by people for different reasons and under different conditions of maintenance. Therefore, every breed requires its own conditions of life, conditions which were involved in its formation.

The more deviations between the biological characteristics of a breed and the conditions of life which are required by it, the less economically advantageous will be such a breed.

For instance, poorly producing dairy cattle, which in their nature cannot give much milk, use good, rich pasture and better feeding with succulent and concentrated foods to smaller economic advantage than a high-producing milch breed. In such instances, the first breed will clearly be lagging economically behind the conditions at its disposal. Such a breed of cattle must be improved sharply by means of hybridization in order to adjust it to the conditions of feeding and upkeep.[62]

[62] Is Lysenko here a Mendelian? The *practical* effect would be the same if Mendelism were true.

On the contrary, highly productive breeds of dairy cattle, falling into conditions of poor feeding and upkeep, will, of course, not give a production corresponding to their breed, but will even survive poorly. In such circumstances it is necessary to adjust immediately the conditions of feeding and maintenance to the breed.

Our zoötechnical science and practice, proceeding from the State plan for obtaining livestock production in the desired quantity and quality must construct all its work in accordance with the principle: *To select and improve breeds according to the conditions of feeding, maintenance, and climate, and simultaneously continuous with it, to establish the conditions of feeding and maintenance according to the breeds.*

The selection and matching of pedigreed animals, most suitably fulfilling the goal set, with simultaneous improvement of the conditions of feeding, maintenance, and care of animals adaptable to development in the desired direction, is the basic procedure for the continuous improvement of breeds.

Hybridization is a radical and quick method of producing variation in stock, in a breed of given animals.

With hybridization, the crossing of two breeds, there comes about a union of the two breeds chosen for crossing, developed by people for a long time by means of a creation of different conditions of animal life. But the nature (heredity) of hybrids, particularly in the first generation, is usually unstable and readily yielding in the direction of influence of the conditions of life, of feeding and maintenance.

Therefore it is especially important in hybridization to observe the rules: To select for a given local breed another which can improve it under the local conditions of feeding, maintenance, and climate. At the same time, in hybridization for the development of characteristics and properties which have taken root in the local breed, it is necessary to make sure of the feeding and maintenance specific to the development of new, breed-improving characteristics; otherwise, the desired qualities may not be grafted to the local improving breed, and a part of the good qualities of the local breed may even be lost.

We cited an example of the application of the general principles of Michurin's doctrine to animal husbandry in order to show that Soviet Michurinist genetics, revealing the general rules of development of living bodies, is likewise applicable in the practical solution of important problems in animal husbandry.

Mastery of Michurin's doctrine must be simultaneously a development and deepening of this doctrine, a development of scientific biology. Such precisely must be the growth of cadres of biologists-Michurinists so necessary for the realization of an ever greater and greater scientific assistance to the collective and state farms in their solution of problems set by the Party and the Government. *(Applause)*

(8) THE MICHURIN DOCTRINE—TO THE CADRES OF YOUNG SOVIET BIOLOGISTS

Unfortunately up to the present the teaching of Michurin's doctrine has not been incorporated in our educational institutions. The responsibility for this lies with us Michurinists. But it will not be inaccurate to say that both the Ministry of Agriculture and the Ministry of Higher Education are likewise responsible.

Until now Mendelism-Morganism is taught in the majority of our educational institutions in the department of genetics and selection and in many instances in the department of Darwinism, while Michurin's doctrine and the Michurin movement in science, nurtured by the Bolshevist Party and by Soviet practicality, are kept in the shade in the universities.

This can also be said about the situation concerned with the preparation of young scientists. As an illustration let us cite the following. In an article "On Doctoral Dissertations and the Responsibility of Opponents" published in the journal *University Herald* No. 4 for 1945, Academician P. M. Zhukovski, as chairman of the Expert Biological Commission attached to the University Certification Commission, wrote:

> A critical situation has developed with dissertations on genetics. *Dissertations on genetics in our universities are extremely rare, even isolated.* This is explained by the abnormal relations, obviously hostile, between the adherents of the chromosome theory of heredity and its opponents. If we are to tell the truth, the former are rather afraid of the latter, who are more aggressive in their polemics. It would be better to put an end to such a situation. Neither the Party nor the Government have forbidden the chromosome theory of heredity, and it is freely offered from university platforms. Yet the polemic is permitted to continue (page 30).

First of all, let us note that in his statement P. M. Zhukovski acknowledges that the chromosome theory of heredity is freely offered from university platforms. He is right in his admission. But he is aspiring to something greater: He desires an even greater flowering of Mendelism-Morganism on an even more enlarged scale in the universities. A significant part of Academician Zhukovski's article is devoted especially to that purpose, reflecting his general direction as chairman of the Biological Commission.

It is not amazing therefore that dissertations on genetics in which the dissertant made even a timid attempt to develop some position of Michurinist genetics were hindered in every way by the Expert Commission. But dissertations by Morganists, whom P. M. Zhukovski patronizes, appeared and were confirmed not quite so infrequently, in any case more often than would be in the interest of true science. True, dissertations of this nature, Morganist in their direction, appeared less frequently than P. M. Zhukovski would desire. But there are reasons for this. In recent years young scientists who are able to comprehend philosophical problems understand, as a consequence of Michurinist criticism of Morganism, that the outlook of Morganism is entirely alien to the world outlook of a Soviet citizen. Academician P. M. Zhukovski's position, advising young biologists to pay no attention to the Michurinist criticism of Morganism and to continue to develop Morganism, does not look well in this light.[63]

Soviet biologists act properly when, fearing the Morganist outlook, they refuse to listen to the scholasticism of the chromosome theory.

I. V. Michurin considered that Mendelism ". . . is not consistent with inherent truth in nature, before which no artificial tissue of erroneously understood phenomena can hold out." "It would be desirable," wrote Michurin, "if the impartially thinking observer would stop before my conclusion and would personally check through the truthfulness of real deductions; they appear as a principle which we bequeath to the naturalists of future centuries and millenniums."[64]

(9) FOR A CREATIVE SCIENTIFIC BIOLOGY

I. V. Michurin established the scientific principles for the control of the nature of plants. These principles changed the very method of thought in the solution of biological problems.

The practical control of the development of cultivated plants and domestic animals presumes a knowledge of causal relations. For biological science to have the strength to help more effectively the collective and state farms produce large harvests, high milk yields, and so forth, it must understand complicated biological interrelations and mechanisms of life, and development of plants and animals.

[63] Michurinism triumphed in the agricultural research institutions in 1940 when Vavilov was arrested. It speaks well for the universities that even in a totalitarian state the unpopular but scientifically sound gentics could survive for eight years. It seems to have been completely exterminated in 1948.

[64] I. V. Michurin, *Works,* III, 308-9. This quotation seems to be an instance of Russian modesty.

The scientific solution of practical problems is the most reliable method toward a sound knowledge of the rules of development of living nature.

Biologists have occupied themselves very little with the study of the relationships and natural-historical congruous bonds which exist between separate bodies and separate phenomena, between parts of separate bodies and links of separate phenomena. Meanwhile only these bonds, relationships, and congruous interactions permit a perception of the process of development and nature of biological phenomena.

The scientific basis of studying biological connections is lost when studying living nature detached from practice.

In their investigations Michurinists proceed from the Darwinian theory of development. In itself Darwin's theory is quite insufficient for the solution of the practical problems of socialistic agriculture. Therefore, at the base of modern Soviet agrobiology lies a Darwinism, altered in the light of the Michurin-Williams doctrine and thus transformed into a Soviet creative Darwinism.

As a result of the development of our Soviet, Michurinist-inclined, agrobiological science, certain of the problems of Darwinism appear in a different light. Darwinism is not only cleared of shortcomings and errors, is not only raised to a very high level, but in a significant degree is altered with regard to a number of its positions. From a science primarily *explaining* the past history of the organic world, Darwinism becomes a creative, *operative* means in its systematic mastery of living nature from the viewpoint of practical work.

Our Soviet, Michurinist Darwinism is a creative Darwinism, raising and solving the problems of the theory of evolution in a new way, in the light of Michurin's doctrine.

I am not able in the present report to touch upon many theoretical questions which had and have great practical significance.

I will briefly touch upon only one of them, namely, the question of intraspecific and interspecific relations in living nature.

The need of reconsidering the question of species formation from the point of view of an abrupt transition from quantitative growth into qualitative species differences has appeared and become more intense.

It is necessary to understand that in the historical process the formation of a species is a transition from quantitative variations to qualitative ones. Such a jump is anticipated by the natural life-activity of organic forms as a result of a quantitative accumulation of the influence of perceptions of the specific conditions of life, and this is completely accessible to investigation and control.

Such a comprehension of species formation, corresponding to

the natural conformities with the laws of development, gives into the hands of biologists a powerful means of control of the very processes of life and through it of species formation.

I think we are justified in considering this statement of the problem that in the formation of a new species, in the production of a new species from an old one, there is not involved an accumulation of those quantitative differences by which varieties are usually distinguished within the range of a species. The quantitative accumulations of changes leading to a spasmodic transformation of an old species into a new form are changes of another order.

Species are not an abstraction but real living units (links) in the common biological chain.

Living nature is a biological chain, broken up, as it were, into individual links—species. Therefore it is inaccurate to say that species do not preserve the stability of their qualitative-species definiteness, for any stage. To speak thus means to admit the development of living nature as a horizontal evolution without abrupt changes.

I am upheld in these ideas by the experimental data on the conversion of hard wheat *(durum)* to soft *(vulgare)*.

Let me note that both of these species are acknowledged as good, faultless, independent species by all systematists.

We know that there are no true winter forms among hard wheats, and therefore hard wheat is cultivated only as a spring and not a winter form in all regions with relatively severe winters. Michurinists have mastered a good method for the transformation of spring into winter wheat. It has already been mentioned that many spring wheats have been experimentally transformed into winter forms. But all of this is concerned with the species of soft wheat. When, however, we proceeded to the transformation and retraining of hard wheat into winter wheat, it was found that following a two-three-four-year fall sowing (necessary for transformation of spring into winter forms) *durum* is transferred into *vulgare,* i.e., one species is transformed into another. The *durum* form, the hard, 28-chromosome wheat, is transformed into different varieties of soft 42-chromosome wheat, during which we do not find transitory forms between the *durum* and *vulgare* species. The transformation of one species into another occurs spasmodically.[65]

Thus, we see that the formation of a new species is prepared for in a succession of generations by a species-altering life-activity, in specifically new conditions. In our case the influence

[65] See page 12.

of fall-winter conditions over a period of two-three-four generations of hard wheat is necessary. In such instances it can spasmodically turn into a soft form without any transitory forms between the two species.[66]

I consider it necessary to note that the stimulus for the statement of the problem of a sound theory on the problem of species, the problem of intraspecific and interspecific interrelations of species, has been and remains for me more than a simple eagerness for knowledge or love for bare theorization. The need to take up these theoretical questions was and is brought to me by the work on the solution of doubly practical problems. A clear presentation of the qualitative differences of the intraspecific and interspecific variety of forms was needed for a proper understanding of the intraspecific and interspecific interrelations of species.

In connection with this, the possibility of the solution of such practically important problems as the struggle against weeds in farming, the selection of components for the sowing of grass mixtures, quick and wide forestation in steppe regions, and many other problems has appeared in a new light.

This has led me to a reconsideration of the problem of intraspecific and interspecific conflict and competition, and after deep and many-sided consideration and study of this problem, to a rejection of an intraspecific struggle and mutual assistance of the individuals within a species and the recognition of an interspecific struggle between different species. Unfortunately I have as yet thrown but little light in print upon the theoretical content and practical significance of these problems.

* * * * *

I conclude the report. And so, comrades, as for the theoretical aims in biology, the Soviet biologists consider that Michurinist aims are the only scientific ones. The Weismannists and their followers, denying the heredity of acquired characteristics, are not worthy of any extensive remarks. The future belongs to Michurin. *(Applause)*

V. I. Lenin and I. V. Stalin discovered I. V. Michurin and made his doctrine the property of the Soviet people. By their great fatherly attention to his work they preserved the wonderful Michurin doctrine for biology. The Party and Government and I. V. Stalin personally are constantly looking after the further development of the Michurinist doctrine.[67] For us Soviet biologists there is not a more honorable task than the creative

[66] See page 12.
[67] This is worth emphasizing.

development of Michurin's doctrine and the introduction of the Michurinist method of investigation of the nature of development of living things into all our activity.

Our Academy must foster the development of Michurinist doctrine as taught by the personal example of solicitous interest in I. V. Michurin's activity on the part of our great teachers V. I. Lenin and I. V. Stalin. *(Tremendous applause)*

VI

The Witch Hunt Gets Under Way

The report of Lysenko occupied the first session of the Lenin Academy. During the next seven days, nine more sessions were held in which fifty-six speakers were given the opportunity to express themselves. Evidences of the coercion of geneticists are plentiful. Only seven speakers dared to defend genetics as it is understood in the free world, and several of these tossed out so much of it to the wolves that it is obvious that they were badly frightened and hoped at best to save only a few fragments of their science from the threatened debacle. The Lysenko group, on the contrary, was confident, aggressive, and all set for the kill. The savage way in which they attacked the minority is unprecedented in the history of scientific academies. Vicious descents into personalities were the rule rather than the exception. The geneticists were heckled from the floor, and threatened and bullied in an atmosphere of the utmost hostility. All Party line scientists west of the Iron Curtain would do well to read and digest the official proceedings. There is now no excuse for ignorance as to how scientists are treated by the Communists.

The Russian authorities apparently did not know enough to conceal what occurred. The sessions were given full publicity. *Pravda* devoted nine issues to the meetings and stenographic reports of the more important speeches were published. Even the minor talks of the "me too" variety were reported in condensed versions. In all, the *Pravda* articles came to 80,000 words compared to the 200,000 words of the official *Proceedings,* which was issued in due course in an edition of 200,000 copies.

The *Proceedings* has also been printed in English by the Foreign Language Publishing House in Moscow. This volume should be in every biological library and it should be obtained before the Russians learn how foolish it makes them appear. An earlier *Proceeding* (English edition, 1936) has been suppressed and is now extremely difficult to consult. The publication of the 1949 *Proceedings* was also a major error. It is a book the Communists should never be allowed to forget. Let us express the hope that they will awaken too late to suppress it effectively.

Twenty-three of the speeches made at the sessions of the Lenin Academy are included in this and the following two chapters. They are translated from the *Pravda* coverage of the meetings and are at times somewhat shortened. In no case has anything been left out which appeared in *Pravda;* all the deletions were done by the Russians themselves. The only selection done by the editor was to choose which of the speeches were to be published. The editor believes that nothing of importance has been omitted.[1] Without exception, the speeches not included were made by the more scientifically illiterate of Lysenko's followers and formed only a highly repetitive Amen Chorus. The editors of *Pravda* had already indicated their estimate of these speeches by the drastic way they had cut them. Still, these unimportant speeches composed of the same cant phrases and stereotyped expressions could, by their very number, set the tone of the meetings.

No speech given in the second session held on the morning of August 2 is worth inclusion. The eight speakers were unanimous in their support of Lysenko and the applause they received helped to show how overwhelming was the hostility to genetics. None of the speakers cited any scientific data whatever or described the evidence upon which his conclusions were based. In the third session held on the evening of the same day, seven more speeches were made. Of them, two are worthy of comment and are reprinted here.

1. S. G. Petrov (§ 1) climbs upon the bandwagon, admits his former errors, and becomes the first of the geneticists to recant. His conversion made him indulge in a little more violence than the *Pravda* account indicates. (From *Pravda,* August 6.)

2. I. A. Rapoport (§ 2) gave the first defense of the gene and of the chromosome theory of heredity. The defense itself was halfhearted and a very obvious attempt was made to give no offense to anyone. Weismann was discarded and denounced and, in addition, the author genuflected before Michurin. His conciliatory efforts, however, were not appreciated. (From *Pravda,* August 7.)

Of the six speeches of the fourth session held on the morning of August 3, only the one made by Professor Z. Y. Beletski (§ 3) contains material of value. He tells us how genetics was liquidated in the Moscow State University. Comrade Beletski undoubtedly told more than he intended. From his attack on genetics we learn how the Department of Biology of the University of Moscow

[1] Abstracts of all of the speeches have been published in *Plant Breeding Abstracts* 18:642-64. 1948. The Russian edition of the *Proceedings* was reviewed in detail by I. Laptev (*Pravda,* Sept. 11, 1948). This review was Englished and printed in *Soviet Press Translations* 3:686-94. 1948.

struggled bravely to preserve the scientific integrity of its members. It was not until their opponents, the personnel of the Department of Dialectical and Historical Materialism, went outside of the University for help that they were beaten, and the "Idols of the Theatre" prevailed in the University. (From *Pravda,* August 7.)

Six speeches were also made at the fifth session but only two are worth reprinting here. The first of these, by V. A. Shaumyan (§ 4), adds a hilarious touch. Comrade Shaumyan is one of Lysenko's scientific illiterates, and his research which proves that a cow gets her teats pulled between six and seven million times in her life, and that this is her main source of exercise, is well worth reading. His conclusions equal those of Lamarck at his funniest. (From *Pravda,* August 7.)

The other speech made by M. B. Mitin (§ 5) is a very different contribution. Mitin is perhaps the most intelligent of Lysenko's supporters. He has a brilliant and subtle mind and a sense of humor which he indulges somewhat rashly. He is well informed and understands thoroughly what he is doing. He does not merely follow his stupid and ignorant colleagues; he urges them on into new absurdities, well aware of the fact that they are destroying Russian science. When his eloquent misrepresentations are wildly applauded, his secret enjoyment becomes apparent. This can be recognized by anyone who knows genetics if he remembers that Mitin is not deceiving himself. If we imagine while we are reading Mitin's paper that it was written by a geneticist, say Dobzhansky, Muller or Sturtevant, we shall have no difficulty in picturing Mitin's private amusement. (From *Pravda,* August 8.)

1

Professor S. G. PETROV

Scientific Research Institute of Aviculture

The wide use of natural plant material for selection in the work of I. V. Michurin was of essential moment. It was this procedure which was adapted by the eminent zoötechnologist M. F. Ivanov in the breeding of the famous Askaniski merino sheep and other animal breeds. Soviet aviculturists have begun to employ widely these methods in work on the creation of meaty types of domestic fowl. As is well known, up until the war the selection in hens in our Country led mainly in the direction of egg production, but now selectionist-aviculturists are trying to develop a meaty type of hen, possessing a delicate meat with excellent tasting qualities. Such a breed of hen is at present strongly propagated near Moscow, and shortly Muscovites will have weighty, meaty, tasty broilers.

What sort of methods are at the disposal of Soviet selectionist-aviculturists? Of course, until recently the theoretical aims of Mendelism-Morganism were strongly professed, even in our practical scientific activity. Many will remember through and through the formalistic book of the well-known aviculturist Mikhalkov. It is impossible not to mention how Academician A. S. Serebrovski tried with all means to direct selection work in poultry-raising along the path of closely related propagation. Inbreeding was used to insensibility.

A series of inaccurate examples was also included in my book, published in 1934. Tribute was likewise given in it to formal-genetic aims.

However, the appropriation of the materialistic method by biology, a thoughtful evaluation of the results of the work of our best selectionist-zoötechnologists, and discussions on the theoretical problems of genetics—all of these were a positive influence on the activity of selectionist-aviculturists. We realize our former mistakes and evaluate now, according to worth, the guidance of M. F. Ivanov on the great influence shown by the environment and, particularly, by feeding, in the creation of a variety. This deeper theoretical thought helps to explain the continuity of the organism with the external environment and the influence of the external conditions on heredity, on the creation of a variety. Thus the Soviet aviculturists approached to Michurinist aims, which proved to be very fruitful.

The proper training of cadres appears as the most important task in overcoming the theoretical errors of Mendelism-Morganism in biology. The existing textbooks on genetics are unfit, they are full of formalistic, antiscientific lies. It is necessary to publish new textbooks, composed on the basis of Michurin genetics. *(Applause)*

2

I. A. RAPOPORT

Doctor of Biological Sciences, Institute of Cytology, Academy of Sciences, USSR

The President of the Academy has given a report on a very general program subject. We must therefore concern ourselves very seriously with Comrade Lysenko's criticism of certain branches of Soviet biological science, particularly the general theory of evolution and the theory of heredity, i.e., modern genetics.

It is entirely natural that different hypotheses which originate in the mind of an experimenter and those theories which have been postulated in the broad field of science are frequently associated with inconsistencies. Truth is born through trial. Thus, the modern theory of light appears as a result of the struggle between two theories developed successively, when victory came first to one, then to the other viewpoint, and the assumption of too rigid an attitude toward either of these theories would have brought harm to science. We in Soviet science must stand apart from the suppression of any point of view which may appear fruitful.

The basis of genetics, as indicated by the term itself, is the gene, the material bearer of heredity. And basically the theoretical disagreement which has developed over this question concerns the gene. The founder of the modern theory of the gene is Charles Darwin. It is sufficient to read several chapters of his book *The Origin of Species*. Without the acceptance of the gene as a material unity, the theory of natural selection could not, of course, exist.

Comrade Rapoport says further that the bourgeois geneticists Bateson, Johannsen, and Lotsy completely rejected the relationship of the gene to the chromosomes. But the idealist-geneticists had to substitute something in place of the materialistic theory which they denied. That is why they explained heredity, for instance, by psychic factors. This sort of theory belongs to a succession of scientists—Semon and others. It is peculiar to many idealists of the type of Driesch, many Lamarckists of the type of Kopp, and others who follow the tenets of Lamarckism. A theory of memory and requirement is accepted by Professor Mach, an idealist who occupied himself with problems of heredity and even did experiments on many animals. He said that heredity can be explained only by requirements appearing outside of matter, with which the new Academician Prezent is in agreement.

The genes and the chromosomes became known as a result of persistent and difficult experiments.

Soviet geneticists have shown that the genes are not unalterable but, on the contrary, are capable of producing mutations. Our geneticists have found chemical means arbitrarily permitting the obtaining of sudden hereditary changes in organisms. There are chemical compounds which produce hereditary changes in every interpenetrating fungus cell.

As a result of this work it can be pointed out that we have completely disproved Weismann's position that the germ cells are enclosed in a special container. There is no container because germ cells are changed with the same frequency as somatic cells. Therefore we are in a position to limit the material bases of life, in an active position to make genes such as they ought to be. There is no container also because embryology has clearly shown that sex cells are not different from somatic.

At present we are on the brink of great discoveries in the field of genetics. Many will recall the fact of the discovery of the existence of the phages—finest of viruses, parasitic on bacteria. The tremendous development of microscopic technique permits us to see phages, destructive provokers of disease in both man and domestic animals. It is possible to see the most minute structure of the phages, to see how they penetrate a bacterial cell, multiply, rupture its capsule, and cause its destruction. The gene is a unity, as yet remote from the possibility of direct visual demonstration already realized in connection with the phages, but in every instance a unity of matter, in work over which there is the possibility to come to great practical progress.

We have to distinguish Soviet genetics from bourgeois genetics. Soviet geneticists have never maintained inaccurate, anti-Darwinist positions. They have bound natural selection into one great principle, which reasonably explained the phenomena of development in organic life and the material substratum of heredity—the chromosomes.

Genetics has described certain mechanisms for producing directed changes within a known measure, one which can be obtained with a repetition of a definite experimental procedure. Thanks to this, genetics can serve our socialistic agricultural economy. It will be of help to employ the method of heterosis over large areas planted in corn, which, regardless of the decree of the February Plenum of the Central Committee of the Communist Party of the Soviet Union (Bolsheviks), is not properly applied. The method given helps to increase corn production not less than 25%. This is not fiction but a concrete fact, and the method indicated must be used by us. This method can be expanded to an entire range of other plants: sugar cane, castor plants, and so forth. The method of heterosis helps to produce a strengthened output of albumens, fats, and carbohydrates necessary to our national economy.

The method of artificial polyploidy, associated with the use of colchicine and producing a doubling of the unities of heredity, is distinguished by unusual prospects. As yet this method has not been thoroughly utilized by us. It is possible to see kok-saghyz, tau-saghyz, the sunflower, hemp, buckwheat, and tens of other agricultural crops, surpassing twofold in their dimensions the diploid plants. The method of polyploidy conceals great practical possibilities, and its theoretical significance is great. In its example it is clear that it is possible for human hands to reproduce species created in nature in the course of many years (tobacco, plum).

Genetics can be of outstanding service to veterinary microbiology in that it helps to produce species of bacteria through disturbance of the pathogenic system. We can produce species which will not bring about disease phenomena but will induce immunity ("living vaccines"). This has been accomplished by many scientists who have given the years of their labors for preservation from tuberculosis, yellow fever, tularemia, and a series of other terrible diseases. But it was accomplished by other methods. At present possibilities of this sort are considerably greater. Then there were examples of accidental findings. Now microbiology, if it will critically accept the fundamental nucleus which exists in genetics, will place it in the service of the needs of our socialistic society.

The principle of natural selection is incompatible with Lamarckism; it contradicts the latter. Lamarckism in the form in which it has been disproved by Darwin and accepted by T. D. Lysenko is a conception which leads to errors. We have been convinced by tens of thousands of exact experiments that alteration of animals and plants cannot be achieved as a result only of our desires. It is necessary to know the mechanisms found at the base of altered morphological and physiological properties. Only through a knowledge of these mechanisms can we achieve alteration of living beings. Even Michurin repeatedly indicated that it is impossible to restrict ourselves only with training in its broad meaning, but that it is necessary to avail ourselves also of the more active methods—to make use of selection and hybridization. And the entire body of Soviet biologists adheres to the theory of selection which Michurin employed in all of his work.

Michurin definitely indicated the possibility of a wide application of genetics not only in orchard but also in field cultivation; he obliged the youth to study genetics. We must carefully nourish the new seedlings in order to raise the new cadres which will be able to advance science forward.

We must give an account as to the accuracy of the principles which we accumulate for our practical activity, and we must not be afraid of criticism, not be afraid of admitting mistakes, and not maintain a course of indiscriminate overemphasis of our achievements or overestimation of what we have.

Only on a basis of criticism of our own errors can we progress further in science, for which our Fatherland calls. (*Applause*)

3

Z. Y. BELETSKI

Professor, Moscow State University

Discussion of Academician Lysenko's report is an event of great importance. The sum total of the controversy in biology, stretching over many years between two trains of thought—between the formal geneticist on one side, and Michurinists on the other—has led to the present session. The events taking place in biology to a great degree recall events which had their place in philosophy. As in philosophy, so at present in biology we encounter phenomena of one and the same form. The struggle is between two motives—the bourgeois-idealistic, and our dialectic-materialistic.

Professor Beletski pointed out that the representatives of the Weismann concept not only defend a bourgeois, theoretical concept in the field of biology but they also drag in harmful ideas on the unity of bourgeois and our science. At present there is current in our press a bourgeois point of view that our Marxist world-outlook, our theory, did not originate from conditions of new social-material relationships of peoples, but that it developed primarily and chiefly as a result of a generalization of all preceding ideological achievement. This bourgeois concept must needs admit that the sciences developed outside of politics, outside of the class struggle, aiming toward the abstract task of a knowledge of the world in general. Hence it is concluded that the true scientist, the modern creator of the social world, is only he who has assimilated all the attainments of both the past and contemporary bourgeois theory. Only one thing must be of interest to such a scientist, namely, how ideas are related and how one develops from another. Therefore, such a scientist can work in the quietude of his study. He is an "observer" of science. Not for him the realities, that conform or not to his theoretical creations of life. To him the theory is important, but not life or practice. The motto of this type of scientist is: Let life adapt itself to science, and if it cannot adapt itself, then so much the worse for it.

This is why our Soviet Weismannists of the type of Schmalhausen, Yudintsev, Alikhanyan, Zhebrak, and others, having assimilated the wisdom of Morgan-Mendelist genetics, decided that they are true scientists, and that our Soviet practice must conform to them. And if it (the practice) does not confirm their theories, so much the worse for it. Here is why over a period of many years they behaved toward the practical progress of Michurinist biology with such unconstraint and contempt. This is why the faculty of biology at the University of Moscow, being the bulwark of Morgan-Mendelist reactionary genetics in our Country, carried on a fierce struggle against the new, true-scientific biology founded by I. V. Michurin and so brilliantly continued and developed in our time by T. D. Lysenko.

I will mention several facts to show what the Weismannists on the faculty of biology at the Moscow State University are doing.

During the span of the past decade, assemblies, scientific meetings, and conferences have been conducted by the Moscow State University's biological faculty devoted to the consideration of one problem—criticism of the theoretical views of Academician Lysenko. It should not be thought that the criticism of Academiciam Lysenko's views ever contained, by any sort of measure, a serious scientific character. No indeed. Academician Lysenko's views were rejected at the threshold as ignorant and not having anything in common with "real" university science. Such an opinion of Michurin's and Lysenko's teaching is held by a majority of the professorial-instructing personnel of the faculty of biology, thus for many years have the students of this faculty been taught. Here is an example. In February of this year an All-Union scientific conference was called by the faculty. The conference continued for some weeks. About forty papers were heard. What sort of problems did the scientific conference consider? Perhaps it discussed the progress of biological science in the practice of agriculture, or perhaps it showed the advantage of our biological science in comparison with bourgeois science? No indeed. From the first to the last paper the conference was directed against the teaching of Academician Lysenko and in defense of bourgeois genetics. The scientists of the biological faculty, it appears, projected the undertaking of a refutation of Academician Lysenko's teaching in the category of a most important problem in biological science during 1948.

Just how far the leadership of the faculty of biology did go toward reaching this aim can also be judged by the methods to which they resort. Let me give several examples.

In connection with an interview given by T. D. Lysenko in the *Literary Gazette* on the conflict, the Scientific Council of the faculty of biology held a meeting at which Academician Lysenko's point of view was submitted to severe criticism. Following the meeting of the faculty's Scientific Council, the Department of Dialectic and Historic Materialism of the Moscow State University organized its own meeting for a discussion of the same subject.

What was the reaction of the leadership of the faculty of biology to this meeting of the Department? It began with a request from the Department of Darwinism that the meeting be held jointly. Why? On this score the Department of Darwinism gave the following explanation: "We fear that the Department of Dialectic and Historic Materialism will not be able to reach a conclusion on the problem independently." When it was politely pointed out to the representative of the Department of Darwinism that its viewpoint was known and that the Department of Dialectic and Historic Materialism decided to consider the matter independently, the biologists resorted to the following procedure. The representative of the Department of Darwinism declared: "If you support Academician Lysenko, you will assume the responsibility with all its subsequent consequences. The opinion of the University must be one."

The Department did not comply with the directive, and upset the unity of the University. It expressed its viewpoint in the *Literary Gazette* and in the newspaper *Moscow University*. This step on the part of the Department did not fail to give its results. The warning given by the Department of Darwinism appeared to be brought into action. Now the leadership of the faculty of biology began to request the removal from the University, not only of the teaching of Academician Lysenko, but of the Department of Dialectic and Historic Materialism. Further events developed in the following manner.

At a responsible meeting of the University the Chairman of the Department of Genetics, Docent Alikhanyan, appeared in the name of the faculty with a declaration. "In view of the fact," he said, "that the Department of Dialectic and Historic Materialism of Moscow State University has not been able to cope with its tasks in the field of biology, has proven itself to be theoretically illiterate, I deem it necessary to raise the question of renovating its personnel." Docent Alikhanyan's request obviously was taken under consideration. A commission was quickly created by the Rector for an examination of the faculty philosophy. The commission worked for a period of two months. The Scientific Council of the University, resting on the report of the commission, accepted a decision in the spirit of the requests of Docent Alikhanyan. The decision of the Scientific Council of the University was not casually brought to completion, but through the force of circumstances, depending neither on Docent Alikhanyan nor the commission. How did the leadership of the faculty of biology behave toward the Department during the period of the preparation by the Rector of measures favorable for its "renovation"? The faculty of biology adopted tactics of obstruction toward the Department of Dialectic and Historic Materialism. The seminar pursuits in the course of dialectic and historic materialism in the Department were discontinued by the departmental dean, C. D. Yudintsev. The Dean requested that the Department replace the chairman of the seminar, Comrade Fuhrman, since he was an outspoken adherent of Academician Lysenko's teaching. The Department did not fulfill this request, as a result of which the students did not study during a semester.

A few words about the students in the Department of Biology. Methods of unbelievable suppression are employed in the Department in its attitude toward the students. Criticism in categorical form of the teachings of Michurin and Lysenko is required of the students in the Biology Department. If, however, notwithstanding this, individual students find themselves in disagreement with the Weismannists, they dare not disclose it openly. Some of these students appearing at the Department of Dialectic and Historic Materialism, in order to receive necessary consultations, insistently ask not to reveal their convictions or their names.

The leadership of the Biology Department actively erodes the views of Michurin and Lysenko, not only out of the consciousness of students but also out of the professorate.

From the foregoing we can see how actively the leaders of the Biology Department of the Moscow State University fought against the teachings of I. V. Michurin and T. D. Lysenko. It is incomprehensible why Yudintsev, Alikhanyan, and others are silent now. One of two things—either they have nothing to say, or they think that now one of the alternate discussions is taking place in which they have no business and which does not concern them. They obviously think that by being silent now they will have the opportunity to call their conference at the Moscow State University and have their revenge.

But let us think that their hopes will not come true. Our Party is strong in that it knows for what it fights, and knows under the banner of what ideas and what theories it conquers. The doctrine of I. V. Michurin and T. D. Lysenko has been verified by the practice of socialistic construction. The theoretical foundation of this doctrine is dialectical materialism. Within it, within this doctrine, lies the future. *(Applause.)*

4

V. A. SHAUMYAN

Director, State Pedigreed Breeding Organization of the Kostromsk Breed of Large Cattle

The so-called theory of the inalterability of genes and chromosomes of the sex cell, a theory of the absolute isolation of the sex cell from the remainder of the living organism, is at its root metaphysical and reactionary. This theory tries to transform us, the Soviet people, into passive spectators of nature, indifferently and submissively awaiting its kindness.

On the other hand, I. V. Michurin's doctrine, fruitfully developed by Academician T. D. Lysenko, demands an active purposeful guidance over plant and animal organisms through the influence of external environmental conditions upon them.

It is necessary to take the general offensive against this idealistic and reactionary theory and its adherents in all branches of Soviet biological science.

I will dwell on fundamental principles and methods in breeding the Kostromsk breed of large cattle. In spite of the conclusions of the idealists-geneticists on the impossibility of changing the breed qualities of animals by means of controlled influence of the external conditions and guided by Michurin's doctrine, we attained considerable success. Under the direction of S. I. Shteiman, Laureate of the Stalin Prize, the superior selective nucleus of the Kostromsk breed—the herd at the state farm "Karavayevo"—was improved and perfected over a period of many years.

Over a period of more than twenty years the herd at the state farm existed under circumstances of diverse and plentiful feed. In the best years the consumption of feed reached 6,000 feeding units per foraging cow. The cows received from 1,000 to 2,500 kilograms of concentrated feed per year.

During the entire year of 1940 a maximal record of milk yield of 6,310 kilograms per foraging cow was obtained over the entire Karavayevo herd. The live weight of cows reached 649 kilograms on the average. The best record holder of the state farm, "Poslushnitsa II," showed a record yield of 16,235 kilograms of milk in 387 days with a butter fat content of 3.92%. Over 70 cows, each with a yield of over 8,000 kilograms, were raised in the herd of this state farm.

The average weight of cow's udders (in full size) amounts to 15-18 kilograms and 22-25 kilograms in certain individuals, whereas it equals 0.5-1.5 kilograms in ordinary cows. The total circumference of udders in individual cows amounts to 1.50-1.85 meters.

Experimental sections of Karavayevo cows showed that the digestive organs, heart, liver, lungs, spleen, and others were as a rule one and a half to twice as large in size and weight as in ordinary low-productive cows. Experiments showed that the arterial and venous pressure, the physiological norms of respiration, pulse, and gas exchange in Karavayevo cows are one and a half to twice as high as in ordinary cows. And even as regards such a constant characteristic as body temperature, the majority of these cows showed a temperature higher than the norm by almost a whole degree.

The question arises in the light of these facts: Is it possible to believe the Morganist theory of the inalterability of the sex cell and of the impossibility of the transmission through heredity of characteristics accumulated by means of the influence of external conditions?

When you emphasize such questions, the formal geneticists say: "You know, of course, these changes exist and it is impossible to deny them, they are even impressive, but they are not new, they were present in hidden form in the primary gene stock, and strictly all this is a result of a new combination or shuffle of genes." (Professor Kislovski.)

How are we to explain the existence of such genes in the course of thousands of years, which specify a cow's milk yield to be 15 or more thousands of kilograms of milk per year? How often a thousand years ago could a cow with a yield of 40-50 kilograms of milk per day appear? Why a calf cannot consume this milk, and in those distant times there was no one to milk the cow. The development of the cow's udder is a most characteristic and irrefutable example of the inheritance of characteristics arising under the influence of the action of external factors.

The state farm "Karavayevo" and the better collective-farm breeding farms in our breeding organization, having at their disposal original breed material with yields of 2,500-4,000 kilograms per year did not reach their present-day position at once. It is enough to point out that more than a 100 strokes by a milkmaid's hands are needed to get one liter of milk, which amounts to about 6-7 million strokes over the entire life of a cow with a yield of 6,000 kilograms per year. Can such a vigorous determining factor of action on an udder applied from generation to generation over an expanse of many years remain without result? We attribute no less importance to the factor of the milking process than to feeding because milking is one of the most important methods and means of exercise for a milch cow.[2]

There is one conclusion. At the base of the breed-developing process of the Kostromsk breed, particularly of the Karavayevo herd, lies expert application of the force of the external factors of reaction on the living organism. It is necessary to direct to these factors plentiful feeding, skillful intensive milking, an appropriate raising of animals, and in close connection with all this, a creative choice and matching of mates.

The purposeful influence of the environment over many years produced those quantitative changes which, on being transmitted through heredity, changed the qualities characteristic of the breed in the original form. These changes are not temporary but organic and physiological, and are therefore inherited.

This was the course we took, the course pointed out to us by I. V. Michurin, T. D. Lysenko, M. F. Ivanov, and P. N. Kuleshov. Following this course any animal breeder will undoubtedly attain success. *(Applause)*

[2] This observation deserves attention.

5

Academician M. B. MITIN

Comrades, the current general meeting of the Lenin Academy of Agricultural Sciences is not an ordinary general meeting—it has a special, one may say flatly, a historical significance. At this meeting a summing up is made of the fight, of many years' standing, between two trends in the biological science of our Country. This fight between trends has the most acute, vital importance, and touches upon the fundamental questions of biological science. The question is whether to move biological science creatively forward, and to arm practitioners of agriculture and animal husbandry with mighty, scientific, effective methods for a further uplift of our Socialist agricultural economy; or whether to be occupied with sterile, antiscientific, scholastic "research," which not only offers nothing to our Country and our State, but actually leads astray practitioners of agricultural economy. The question is whether to develop further our Michurinist, consistently materialistic, Soviet trend in science, which has enriched biological theory with discoveries of the greatest scope and significance, and which represents qualitatively a new step forward in the theory of evolution; or whether to follow obsequiously the antiscientific, idealistic conceptions of the foreign, bourgeois "authorities," which radically undermine the theory of evolution.

In tracing through scientific work, literature, practical results, discussions and speeches of the representatives of biological science struggling among themselves, it becomes manifestly possible to determine that in our Country two radically opposite trends had been formed and established. One of them is rightly called Michurinist—after its creator, the great naturalist, reformer of nature, I. V. Michurin; the other trend is the reactionary-idealistic, Mendelist-Morganist trend, whose founders were the bourgeois scientists Weismann, Mendel, and Morgan.

The Michurinist trend bases its methodology on the principles of dialectical materialism; it develops creatively the evolutionary theory of Darwin, rejecting at the same time his one-sided, erroneous, and obsolete theses. This trend is closely tied with life, with the practice of the Socialist agricultural economy; it works successfully on improving the old and creating the new strains of plants and breeds of animals; it fruitfully moves biological science forward; it represents, in the true sense of the word, a *people's* trend, bringing into effect a daily, living tie with the collective farms, with experimental stations, with selectionists, agronomists, and the forward ranks of the collective-farm workers.

The Mendelist-Morganist trend in biology, on the contrary, continues and develops the thoroughly idealistic and metaphysical teaching of Weismann concerning a difference in principle between the immortal, uninterruptedly continuous "substance of inheritance," and the so-called mortal "soma." No matter with what reservations in regard to Weismann's doctrine our representatives of the Mendelist-Morganist trend surrounded their statements, the heart of the matter is that their *theoretical foundation,* their initial theoretical base, is *Weismannism,* this reactionary, completely bankrupt teaching, which excludes active influence of man on directed changes of plant and animal organisms.

Representatives of the Mendelist-Morganist trend have been operating for many years with sterile, cabinet experiments, without any contact with life, with the needs of the people and of the Socialist construction. This is an *anti-people* trend in science.

To what repulsive ugliness, arousing a legitimate sense of indignation in the Soviet people, this trend leads, has been illustrated by T. D. Lysenko in his report when he quoted, as an example, the research of Comrade Dubinin concerning the effect of the Great Fatherland War on the chromosomal apparatus of the fruit flies.

Dubinin deserves to become a common noun to characterize loss of contact between science and life; to characterize antiscientific, theoretical research; to characterize the pseudoscientific nature of the Mendelist-Morganist formal genetics, which impels toward this kind of "research."

Mendelism-Morganism, as a definite bourgeois trend in biological science, arose in Western Europe and America at the end of the last and the beginning of our century.

Representatives of the Mendelist-Morganist trend—Morgan, Johannsen, De Vries and others—directed all the deductions of their research against Darwin,[3] against his evolutionary teaching, against the theory of natural selection. Further spread of Mendelism-Morganism serves as a patent confirmation that this trend in biology is directed pointedly against the theory of evolution, against the very idea of development of nature.

It is a known fact that one of the active propagandists of Mendelism-Morganism in our Country at the end of the twenties was Professor Yu. A. Philipchenko. He wrote as follows:

[3] This statement is false, as any examination of genetic literature will show.

The doctrine of mutability and the entire contemporary genetics, of which it is a part, are far from being tied inseparably to the doctrine of evolution. . . . A geneticist can calmly go on working in his field, without even giving a thought to evolution. . . . Quite conceivable is the position . . . of a geneticist who is a profound agnostic in the questions of evolution.[4] (Yu. A. Philipchenko, *Mutability and Methods of Its Study*, pp. 249-50, 1929, 4th edition.)

As another ardent defender of Weismannism and autogenesis, we had also Professor N. K. Koltsov (a eugenicist and advocate of racial theories in biology). Starting with the theory of autogenesis and "pure hereditary lines," Professor N. K. Koltsov offered in his writings, under the banner of science, some most reactionary and insane delirium. Thus, he wrote: "Those who made Europe's history belong to a few hereditary lines, and these lines are closely interconnected by blood kinship." (N. K. Koltsov, "Genealogy of C. Darwin and I. F. Galton," *Russkyi Evgenicheskyi Zhurnal,* Vol. I, No. 1, GIZ, 1922, p. 69.)

The theory of Weismann-Mendel-Morgan also found expression in the works of Professor A. S. Serebrovski, Professor N. P. Dubinin, Professor A. R. Zhebrak, and is now being worked out in the works of Academician I. I. Schmalhausen. The "works" of Academician I. I. Schmalhausen are at present the central works in our Country, representing and expressing Mendelism-Morganism at the present stage.

Academician I. I. Schmalhausen, in his work *Factors of Evolution* (1946), writes:

Mutation is always a new acquisition of the organism, while modification is a sort of a superstructure—a variation of an already existing organization. Mutation is transmitted to progeny according to strictly ordered laws. These laws were discovered by G. Mendel. So far as their essence is concerned, they have been confirmed and subjected to a very thorough analysis in contemporary genetics (especially by the T. H. Morgan school) (p. 13).

Thus we see here that Academician I. I. Schmalhausen refers to Mendel and Morgan as the chief authorities who have discovered the basic laws of mutational changes.

In spite of numerous reservations which can be found in the books by I. I. Schmalhausen, his conception, his basic point of view, expounded in a series of his works and particularly in the book *Factors of Evolution,* reproduces the autogenetic, Weismannist conception in biology.

As the most important concept, drawn to explain the funda-

[4] Quotations out of context.

mentals of evolution, he establishes the concept of a "substratum of phylogenesis." Since the task of phylogeny, as is well known, is the discovery of laws of nature, according to which different species of organisms arise and develop, then naturally, it is precisely the "substratum of phylogenesis" which is the basic carrier of heredity.

Let us quote some statements of the author of the book *Factors of Evolution*. Academician I. I. Schmalhausen writes:

> . . . *nuclear structures* are the specific *substratum of phylogenesis*, in which are fixed all hereditary changes, i.e., all the changes in the norm of reactions, among them also the changes in ontogenesis, changes in organization and its characters, and changes in adaptive reactions (modifications) of an individual organism. (*Ibid.*, p. 74.)

Thus, "substratum of phylogenesis" contains everything: hereditary changes and modifications and changes of ontogenesis and mutation.

"Substratum of phylogenesis" is a metaphysical and scholastic concept. In effect, it is but another expression of the Weismannist "substance of heredity," a repetition of the old, long-exposed by us, reactionary ideas of "fund of genes" that representatives of formal genetics made so much of.

In full accord with Weismannism, I. I. Schmalhausen denies any substantial importance of external factors in the evolution of organic forms.

"The external factor," writes Academician I. I. Schmalhausen, "gives but the first impetus, upon reaching the threshold of reactivity of the tissues of the organism, an impetus that activates the internal mechanism of a definite complex of form-organizing processes. It does not determine either the quality or scope of reaction. At best (and even then not always), the external factor merely determines the time, and sometimes the place of its realization." (*Ibid.*, p. 82.)

Thus, in the opinion of Academician I. I. Schmalhausen, the external factor in evolution is not a cause; it determines neither the quality nor the scope of the reaction of the organism to the environment. Everything is stored up in the so-called "substratum of phylogenesis."

Such are the positions occupied by Academician I. I. Schmalhausen, expounded in his book *Factors of Evolution*. They speak quite clearly that in his basic concepts, in spite of particular reservations, the fundamental, theoretical, initial base, the fundamental position of Academician I. I. Schmalhausen is completely clear: this position is essentially the position of Weismannism.

Academician I. I. Schmalhausen develops further—in his book

Factors of Evolution and in his other works—harmful, anti-scientific theses about the correlation between the "wild" forms of the organic world and the "cultured" forms. He considers that in the wild forms there existed unexposed, unrevealed, or, as he writes, "reserve mutations," "reserves of mutational changes." Formation of strains and formation of breeds is nothing but a disclosure of these reserves of mutations, which were stored a priori in the wild forms. It follows thus that the cultivated species of the plant and animal world are not actually products of culture, products of tremendous, productive work of generations of people, of theoreticians and practitioners of agriculture and husbandry; but are merely the result of a "revelation" of a reserve of mutability, previously stored (one might ask in this case—by whom?) in the "wild" forms.

Academician I. I. Schmalhausen, characterizing the general process of evolution, says that gradually there occurs a general reduction in the "reserve of hereditary mutability in the population." "This process," he writes, "of the loss of evolutionary plasticity of forms I call 'immobilization.' "

"Immobilization," he writes further, "occurs also upon standardization of strains and breeds cultivated by man."

Thus, according to I. I. Schmalhausen, in breeds and strains cultivated by man, which are useful and necessary to man and, thanks to that, have become standardized, there occurs a "loss of evolutionary plasticity," "immobilization." In other words, the plant strains and cattle breeds deteriorate,[5] change for the worse, lose their "wild" force and "charm." What shall we call such "theories" and such "science"? In essence, it is a limiting theory which interferes with the further development of the Socialist agricultural economy and is only capable, if not repulsed, of demoralizing and disorganizing the contingents of our agricultural economy.

In the book by Academician I. I. Schmalhausen, *Factors of Evolution,* one may even find references to dialectical materialism, but these are mere words while, in effect, the methodology on which the book is built has nothing in common with dialectical materialism. This book is metaphysical and idealistic. The methodological basis of the author's conception is the familiar theory of equilibrium.

Let us cite, for instance, the following statement of Academician I. I. Schmalhausen. He writes: "If we speak of the nucleus and its chromosomes as a system ('the balance of chromosomes' and the genic balance), then we have to admit that it is in a state

[5] Note the change in meaning of the statement couched "in other words."

of *little mobility, yet at the same time of a relatively not too stable equilibrium*" (page 75).

Thus we see that the author applies all the basic categories (stable and unstable equilibrium, etc.) of the Bogdanov-Bukharin theory of equilibrium.

To counterbalance the formally genetic, reactionary, idealistic trend in biology, there grew up in our Country, and got stronger, and was given a rich development, the Michurinist trend in the biological science, whose leader and chief representative in our Country is Academician T. D. Lysenko.

Let me remind you that K. A. Timiryazev, speaking of the perspectives of development of Darwinism, considered that the further stage, the higher step in the development of Darwinism will be the discovery of laws and methods with the aid of which it would be possible, as he put it, "to sculpture organic forms." What Timiryazev used to dream about was realized by I. V. Michurin. He has discovered these new laws of development of life and has worked out the methods of a directed change of the nature of plants. Knowing the laws of plant development, I. V. Michurin, like a genius, "sculptured" new organic forms, and at that, such forms are necessary and useful to man.

Darwin, as is well known, to use Engels' expression, "is diverted away from the *causes* that evoke changes in different species." (K. Marx and F. Engels, *Works*, Vol. XIV, p. 70.)

Whereas Michurin, leaning upon all the wealth of Darwin's evolutionary teaching, had studied the causes of individual changes of organisms, had discovered the laws of governing the development of plants, and had worked out methods of evoking purposive[6] changes.

The teaching of I. V. Michurin, creatively developing the materialistic nucleus of Darwinism, is at the same time profoundly dialectical. I. V. Michurin started with the hypothesis that the historical past of an organism is the foundation, the basis on which alone can develop its present and its future. I. V. Michurin proceeded from the unity of phylogenesis and ontogenesis. He has established the correct interrelationship between the historical past of an organism and its hereditary basis.

I. V. Michurin has fully censured and overcome the breach in principle between phenotype and genotype, between ontogenesis and phylogenesis, so characteristic of Mendelism-Morganism.[7]

I. V. Michurin looked at the organism as indissolubly connected with its external environment. He has clarified the tremendous importance of the external environment in the forma-

[6] Is purposiveness materialistic? Or is Mitin having fun at the expense of commissars he considers stupid?

[7] See note 25, page 45.

tion of the organism, and has analyzed the various stages of this formation.

I. V. Michurin, in his notes *External Environment,* wrote:

> Apparently, some people who consider themselves scientific connoisseurs of the laws of the plant kingdom naïvely regard as doubtful my assertion about the influence of the external environment on the process of formation of new forms and species, as though as yet not proved by science.
>
> . . . Thinking of these would-be scientists, one does not know what to wonder at more: their extreme short-sightedness, or their thorough ignorance and absence of any sense in their world-view.
>
> First of all, it would be interesting to know—do they really think that all the 300,000 different species of plants were created (outside of any influence of the external environment) solely by means of hereditary transmission of the properties of their progenitors? But such a judgment would be complete absurdity. One cannot, after all, really suppose that from the first engendered individuals of the living plant organisms, by means of their cross-fertilization, gradually, in the course of tens of millions years, was created the whole, now-existing plant kingdom all over the globe, without any participating influence of the external environment, the conditions of which had so often and so sharply changed in the course of the past ages and millennia. . . .[8] (I. V. Michurin, Vol. III, p. 255.)

I. V. Michurin returns to the problem of the role of the external environment in several other works. Thus, in his article "Bureaucracy in Science," I. V. Michurin writes:

> Only through joint influence of the hereditary transmission of ancestral properties and the influence of the factors of the external environment, were created and continue to be created all the forms of living organisms. It is impossible to object to this incontestable truth. (I. V. Michurin, *Works*, Vol. I, p. 483.)

Thus did I. V. Michurin pose the problem of the role played by the external environment. As we see, I. V. Michurin, as a materialist-dialectician, in his approach to the problem of how to control plant organisms, combines the historical past of the development of organisms with the role played by the external environment.

I. V. Michurin takes into account the historical path of development of organisms, as well as how, along this path, the organism would become adapted to the conditions of its existence. The theory of mentor, which he had worked out, belongs among outstanding scientific achievements.

[8] Biologists, of course, look upon the environment as the selecting agent. Fitness consists of adaptation to environment. This concept seems to have been a little too complex for Michurin to understand.

When touching upon the question of the external environment, it is necessary to pause briefly on the problem of Lamarck.

Mendelists-Morganists have turned Lamarck into a bugaboo, a term of abuse. It is enough to say, "This is Lamarckism," to make them begin to repudiate it, like a devil confronted with Holy Water. Whereas the real, scientific truth about Lamarck is this:

Lamarck, first in the history of the development of science, has established the thesis that the correct classification of organisms is a reflection of the order and development of some organisms from others. He has pointed out the decisive influence of the external environment on the development of organisms, starting from the point that not the form determines the functions of organisms but on the contrary, the functions, directed by the influence of the external environment, determine the form.

Lamarck's teaching, as is well known, arose in connection with the ideas of the French Enlightenment and French materialists. It reflected the revolutionary epoch of the time. His teaching was saturated with philosophic content, and was marked by its materialistic character. Reaction against the French Revolution also caused a strong reaction against the ideas of Lamarck, which lasted throughout the nineteenth century.

K. A. Timiryazev used to say that sober Darwinism alone allots Lamarckism its rightful place in science.

We must note that Darwin regarded it as a serious fault in himself that he had underestimated the influence of external environment on organism. In his letter to Wagner (1876), he wrote: "My greatest error is that I have insufficiently evaluated the direct effect of environment on organism, i.e., the effect of climate, food, etc., independently from the effect of natural selection."

Darwin gave a scientific explanation of the evolutionary process. In this lies his immortal achievement for science, for humanity. But, after all, many decades have passed since Darwin. Science and life have accumulated a tremendous quantity of new facts and phenomena. People who call themselves "orthodox Darwinists," like B. M. Zavadovski, apparently want to say that they follow Darwin's teachings without any amendments, i.e., that they follow also his errors (elements of Malthusianism, gradual evolution, refusal to recognize saltatory changes, etc.), his obsolete theses, and do not want to advance forward. Let them, then, remain on this "general" line. Life and science will bypass them, moving forward.[9]

[9] Geneticists here are accused of not deviating from Darwinism and earlier in this essay of being anti-Darwin (see note 3). Mitin is a humorist.

The Michurinist trend in biology represents a qualitatively new, higher step in the development of Darwinism. I. V. Michurin, having mastered all the wealth of Darwinism, all that was the best in Darwin's teaching, took a tremendous, creative stride in the development of Darwin's teaching.

Darwin laid the foundations of a scientific explanation of evolution, whereas I. V. Michurin *created* evolution. Precisely in this is that new element, the gigantic stride forward in the development of Darwinism, which characterizes the Michurinist trend.

We ought to be proud that our countryman—the great scientist I. V. Michurin—discovered a qualitatively new stage in the development of biological science, blazed new trails in it.

We may be proud that our Soviet scientist, I. V. Michurin, discovered and mastered the laws of deliberate control of the development of organisms. Let all sorts of cosmopolitans in science assert that "questions of priority are of no importance in science." As for us, we shall be filled with legitimate pride that this greatest contribution to the biological science belongs to a Soviet scientist.

The Michurinist trend in biology opens limitless perspectives before the biological science, particularly under the conditions of our Soviet land, under conditions of the collective-farm system. Knowing the laws of changes of heredity of organisms, and skillfully utilizing them, our scientists will work successfully over the improvement of the existing plant forms and animal breeds, and on creating new ones, needed by a Socialist country, by Communism.

Michurin used to dream how to turn our country into a flowering garden. This dream is being successfully transformed into actuality in the land of Stalin's five-year-plans.

The teaching of I. V. Michurin is a shining example of application in scientific research of the method of materialistic dialectics. I. V. Michurin, himself, valued greatly the philosophy of dialectical materialism.

Listen to his inspired words about Party adherence to the philosophy of dialectical materialism:

> Party-adherence in philosophy represents its basic, orienting factor. The order of things determines the order of ideas. A progressive class, such as the proletariat had shown itself to be, is also the carrier of a more progressive ideology; it forges a single, consistent, Marxist philosophy.

And farther on:

> Only on the basis of the teaching of Marx, Engels, Lenin, and Stalin,

is it possible fully to reconstruct science. The objective world—nature—is primary; man is a part of nature; but he must not merely contemplate this nature externally, but, as Karl Marx had said, he can change nature. The philosophy of dialectical materialism is a weapon for changing this objective world; it teaches to influence this nature actively and to change it; but to change and influence nature actively and consistently is within the power of the proletariat alone—thus says the teaching of Marx, Engels, Lenin, and Stalin—the unsurpassed giant-minds. (I. V. Michurin, Vol. I, p. 447.)

Thus did I. V. Michurin tie his doctrine with the philosophy of dialectical materialism. This is how deeply he saw the inner connection existing between his *efficient* teaching of controlled development of plant organisms, and the *efficient* philosophy of dialectical materialism.

Only that science may be called true science which leads toward discovery of the laws of development of the world surrounding man, which opens to man new horizons and perspectives. Only that science may be called true science which serves practice, is verified in practice, gives tangible results in practice. All this is characteristic of the Michurinist trend in science.

The future belongs to the Michurinist trend. V. I. Lenin and I. V. Stalin have given the highest evaluation of I. V. Michurin's theoretical works and practical results, when they called him the great reformer of nature. This is why the Michurinist trend in science is in all ways supported and fostered by the Soviet Government, by our great Party of Bolsheviks, and, personally, by Comrade Stalin.

The followers of Mendelism-Morganism have been warned more than once that their trend in biology is alien to Soviet science, that it leads to a dead end. In 1931, the Party condemned the Menshevist idealism in philosophy and natural sciences. This had a direct bearing on such people as Professors Serebrovski, Dubinin, and others. In the decisions of the public and party organizations of the Institute of Philosophy and Natural Sciences, the following gravest errors on the part of several people in the field of biology and physics were pointed out: "going over to positions of autogenesis," "Machistic opinions in the realm of physics and mathematics," etc. ("For a Turnabout on the Philosophical Front," 1931, published by *Moscow Workman,* p. 231).

In 1939, during the broad discussion of genetics, organized by the journal *Under the Banner of Marxism,* the Mendelist-Morganists were decisively rebuffed; there was criticism of the theory of the gene, of the loss of contact between the formal geneticists, and practice, etc. Nevertheless, the Mendelist-Morganists not only failed to draw from this proper lessons, but continued to defend, and to go still farther in, their erroneous views.

Discussions, prolonged beyond all measure, and the active propaganda of their views by Mendelists-Morganists are doing a substantial damage to the cause of ideological education of our contingents.[10] The essential meaning of the present session must consist in finally putting an end to this immeasurably prolonged discussion; in exposing and smashing to the end the antiscientific conceptions of Mendelists-Morganists; and in thus laying a foundation for the further flourishing of Michurinist research, for further successes of the Michurinist trend in biology.

At the current session it is necessary to note the role of Academician T. D. Lysenko in the struggle of the progressive trend in biology against the reactionary trend. At the moment, we do not have the opportunity to touch upon the most fruitful, theoretical, and practical results in the cause of developing the Michurinist trend, which are connected with the name of Academician T. D. Lysenko. We lack the opportunity to discuss in detail his theory of the stage development of plants, which represents a most important theoretical conquest of biology, or discuss his views on the problems of heredity and other most important problems of biology. At the moment, I would like merely to note the following. Boldly and decisively, with his typical firmness and passion, T. D. Lysenko has been exposing and is exposing Mendelism-Morganism. He had to overcome tremendous difficulties, he was subjected to slander, he was denied "scientific quality," a mass of obstacles was placed in his path, but he went ahead courageously, like a true innovator in science, paying no heed to anything; and he defended, fighter-wise, the principles of his positions, he defended the banner of the Michurinist trend.

Academician T. D. Lysenko—the Michurin of today—has made a tremendous contribution to the development of biological science and to the practice of the Socialist agricultural economy. I believe that I will give expression to the opinion of an overwhelming majority of those present, if I shall say that it is thanks to the courageous and fearless fight by Academician T. D. Lysenko against the conservatives of science, that the further development of the Michurinist trend in biology has been achieved, that such considerable successes in our agrobiological science have been achieved. *(Prolonged applause)*

[10] A possible clue as to why genetics was destroyed.

Under the leadership of the Bolshevik Party, in our Country is taking place the majestic process of building of Communism. Communism means a bright, joyful, and not too distant future. Already now, we can measure approximately in years the distance to Communism. On this fighting path of building a Communist society, our scientists, innovators in the realm of science and practice, are entitled to the most honored place.

Under the leadership of the greatest genius of our times, our beloved and dear teacher—Comrade Stalin—the Soviet science and our scientists-innovators will achieve even greater successes! *(Prolonged applause)*

VII

A Forlorn and Futile Stand

The geneticists offered no real opposition to the followers of Lysenko until the latter half of the eighth session, which was held on August 5. Of the eleven who spoke at the two preceding sessions, which met on August 4, only B. M. Zavadovski did not join the hue and cry. Most of the speakers continued the routine denunciations of genetics and they can be omitted here without loss. Two of the Lysenkoists, however, exposed the Communist standards as no enemy of Communism has ever succeeded in doing. In all, nineteen speeches were made during the two days. The eight which are included in this chapter are given in their chronological order.

K. Y. Kostryukova (§ 1) is the true believer *par excellence.* (*Pravda,* August 7.) The Madam Professor illustrates better perhaps than any other the fundamental conflict between science and Communism. Science may be all right in its place, but it had better not conflict with the orthodox ideology. She also illustrates what we might state as a general principle, that science is bound, sooner or later, to conflict with any system which is intellectually closed.

The speech of Academician B. M. Zavadovski (§ 2) discloses a personal tragedy, a tragedy which is more than physical. (*Pravda,* August 11.) Like Mitin he understands the issues, and like Mitin he has had to make a choice between two degrading courses of action, but, unlike Mitin, he tried to rescue some remnant of science from the debacle, even at the cost of his personal integrity.

Portions of genetics had been so thoroughly smeared in Russia that Zavadovski saw they could not be saved. These he threw overboard in an attempt to stave off the complete surrender to quackery. Weismannism, Malthusianism, Mendelism, and "formal genetics" he discarded. He did not lose sight of the fact, however, that Lysenko was an incompetent Lamarckian. He sought to establish the fact that his Communist orthodoxy was just as good as Lysenko's, that he had been slandered by Lysenko's followers, and that his disagreements with Lysenko were personal and scientific and not political. He seemed to believe that if he did not offend the notions of his neighbors too greatly, he might retain some little influence which he could use for the general good.

Zavadovski certainly did not know at this time that the Central Committee had received Lysenko into the Apostolic Succession and that consequently in matters biological Lysenko was incapable of error. Thus Zavadovski failed rather ridiculously. Mitin, on the other hand, preserved his personal dignity, his sense of humor, and his ability to injure the individuals and the system which had penalized his intellectual honesty. Of the two, Mitin probably preserved a larger fraction of his self-respect. Before we condemn either Zavadovski or Mitin too harshly we should ask ourselves what we would do in their place.

The only speech of the seventh session of real interest is the one delivered by Th. A. Dvoryankin (§ 3). He disposed of Zavadovski in a truly Russian manner. (*Pravda,* August 11.)

Eight participated in the session of August 5 and were evenly divided: four defending genetics and four attacking it. Three of the attackers said nothing that had not been said before but the speech by the fourth, A. V. Mikhalevich (§ 4), is well worth reading. (*Pravda,* August 11.) Mikhalevich is evidently an excellent orator, to judge by the reception of his speech, but he would hardly rank as a scientist. He is a deputy editor of the *Ukrainian Pravda,* and a skilled demagogue. Such a speech as he delivered would not be unusual in a political compaign, in fact we expect such speeches from our politicians and do not take them too seriously. In a scientific assembly, however, it produces something of a shock. It may be a foretaste of what we should have to bear if scientists ever become subordinated to politicians.

The speeches of the four geneticists (§ 5, § 6, § 7, § 8) are revealing and deserve a most careful analysis. (*Pravda,* August 9, 10.) They should certainly be investigated thoroughly by some future historian of science. Without exception the speakers yielded in part to the hostile atmosphere and forswore a part of their knowledge. Thus S. I. Alikhanyan (§ 5), using the conventional phrases, discarded Weismann, Johannsen, De Vries, Bateson, Lotsy, Serebrovski, and Philipchenko. I. M. Polyakov (§ 6) went even further. P. M. Zhukovski (§ 7) declared that he had never opposed Michurin and defended his actions in not accepting bad theses which were Michurinistic from students for publication. A. R. Zhebrak (§ 8) neither attacked the quacks nor defended the geneticists, but confined himself to a factual statement of the results of his researches. Nevertheless, each one of the geneticists showed flashes of real courage and struggled manfully to preserve some remnant of his intellectual honesty.

The sequel will not be forgotten. All four recanted: Alikhanyan, Polyakov, and Zhukovski two days after they gave their papers and Zhebrak four days later. For a glimpse of scientists under coercion, we should read their original papers and their recantations published in Chapter X.

1

Professor K. Y. KOSTRYUKOVA

Director, Department of Biology, Kiev Medical Institute

In his letter to the students of the Kapryska school, V. I. Lenin wrote: "In every school it is most important to have an ideological-political orientation of the lectures."

This premise is very clear to us university teachers and in general to all teaching personnel. We consider ourselves deeply responsible for the education of youth, the future builders of Communism. But then, of course, we educate first of all through the ideological content of our lectures. And it is therefore understood what great injury can be done to the cause of the building of Socialism by a lesson that is on an insufficiently ideological level, not even mentioning a frankly reactionary lecture.

Therefore, the facts which were told us yesterday by the Director of the Philosophy Department of the Moscow State University are simply terrible. For a number of years the young biologists at the Moscow State University have been trained in the spirit of reactionary theory.

It is necessary to indicate that the harm brought to the biological collectives of the Moscow State University is not limited only to that university. The biological collective is a great body. From it have been drawn the cadres for the editorial staffs of biological journals, and the critics of biological articles published in those journals. Hence it becomes clear that such biological journals as the *Journal of General Biology, News of the Academy of Sciences, USSR (biological series), Reports of the Academy of Sciences, USSR* (articles in which biological problems were discussed), never published a single article of Michurinist tendency. In this manner there appeared an entirely specific selection of articles, and consequently the propaganda of the Morgan doctrine was far spread through these journals across our vast Fatherland.

The Ministry of Higher Education of the USSR is located in Moscow. The Ministry of Higher Education approves the programs and the textbooks for the whole country. Who reviews these programs and these textbooks? The very same biologists who are in Moscow. The influence of the Moscow biological group is felt everywhere.

I had occasion to come directly in contact with this influence during the war years. In 1942 a program approved by the Committee on Higher Education and the Ministry of Public Health of the USSR was sent to the Kiev Medical Institute, where I am Director of the Department of Biology. This program was so terrible in its ideological content that I immediately wrote a memorandum to the Institute. Concurring, the Institute sent it to the Committee on Higher Education.

It should be mentioned that this program propagated bourgeois genetics. In order to form some idea of this program, I will mention that in the entire program, a program for general biology, the name of the great biologist Michurin, transformer of nature, was not mentioned once.

The reply came very soon. It was suggested that I construct a plan for a program in general biology for medical institutes. The project was prepared in good time and sent away.

And from that time everything became quiet. The years 1943, 1944, 1945 passed; the war ended. There was no program. What did it mean? It meant that the old program continued all the time.

Finally, in 1946 (the program was published in 1945, but we received it in 1946), a new program appeared. This program was even more terrible.

The year 1946 passed. So did 1947. In the spring of 1948 we received a reference from the Ministry of Higher Education. This reference was a small piece of paper on the middle of which was written: On such and such a page insert the name Michurin; on such a page, following certain words, insert the name Schmalhausen, and so forth. No reply has as yet been received to my protest.

It has now become clear to me what was wrong: the program was imposed by certain individuals to whom it was necessary to continue that which was propagandized in the program.

I want to say that all this—the programs, the lectures in accordance with these programs, and the textbooks written in accordance with these programs—bring colossal harm, which is hard for one even to grasp. The subject has been raised more than once from this platform during this session. The fact is that our youth greedily absorbs the knowledge that is taught from the cathedrae of the universities. Sometimes certain ideas which it absorbs enter so deeply that they are retained throughout life. That is why it is very essential to take measures immediately that there shall be no more such facts as those that have been related here.

Yesterday Comrade Rapoport spoke from this platform. He spoke as a real Morganist, a convinced Morganist. Comrade Rapoport defended the Morgan orientation in such a manner that from the first it seemed that everything set in this theory is all right. The gene appeared to him dressed in new, fashionable clothing, biochemical clothing. There was talk already about gene-hormones.

But you must be honest, Comrade Rapoport. You ought to have said that your newly stated hypothesis is an unsubstantiated hypothesis; instead you project it as a definite fact. It must be said that the introduction into their science of such unproved hypotheses as definite facts is very characteristic of the Morgan orientation. It is characteristic of the founder of the theory, Morgan himself. In general, Morgan did not recognize the word "hypothesis"; even the word "theory" was below his merit. All of his fabrications were called "laws." In any textbook of genetics will be found the law of crossing-over, the law of lineal arrangement of genes, and so forth.

I would say that such exposition of unproved hypotheses displays the excessive pride of the advocates of the Morgan line. We have already heard how their pride also shows itself in that they acknowledge themselves as the only true scientists, and therefore do not condescend to a criticism of other theories.

Upon what lies the pride of the advocates of the Morgan doctrine? Their teaching is, as they themselves define it, teaching about the gene. The gene is the central concept of Morgan genetics. What is this gene in the definition of the Morganists? The gene is a material particle. Yesterday Comrade Rapoport told us about it and insisted on it specifically. Well, what sort of a material particle is it? It is a special substance which is the bearer of heredity. Thus, according to the Morganists, there is a specific substance, bearer of heredity, and all the rest has no connection with heredity.

Just consider carefully what is said! Heredity, an attribute of a living body, is torn away from it, is contrasted with it.

In his article in 1936, published in our journal *Nature,* Muller writes very clearly about it. He says that in the cell there is a nucleus, there is protoplasm, but not everything alive in the cell is a bearer of heredity. Only a small part of the cell's substance, the chromosomes, possess this property. True, it is now said that there also are genes in the plasm. However, essentially it does not change the matter. Both the plasma gene and the chromosome gene are particular hereditary substances. Such special fabricated substances and forces are known in the initial development of many sciences. They are called upon to explain phenomena incomprehensible at a given stage of development. Thus, for instance, in physics in order to explain thermal phenomena, thermogen was invented, in chemistry in order to explain combustion, phlogiston. In biology a life force had to explain incomprehensible life phenomena.

Thus, the properties of substances are separated from the body, are contrasted with it, as of no importance. We see the very same thing in genetics: heredity is an attribute of living matter, yet the hereditary substance is separated from and contrasted with it as of no essence. Geneticists study this hypothetical substance. They even connect it with a material substratum, the chromosomes, and see in it a confirmation of its existence. This, however, does not deprive the hereditary substance of those features by which other fabricated matter is distinguished. A hereditary matter, contrasted to living matter like thermogen and phlogiston, does not exist.

In this contrast is shown the absolute dualism characteristic in all idealistic, vitalistic explanations of life. Here is that, it appears, in which the Morganists take pride. The gene is pure fiction, if you did not believe, Comrade Rapoport, that it is a material particle. The electronic microscope will not save you. You may see in the electronic microscope various, favorable, minute particles, and these may be particles of chromosomes, but you will not see a gene, because it does not exist, just as a vital force does not exist.

Thus, it appears that knowledge about the gene is in a prescientific period of its development. The theory of the gene is a false theory, retarding the development of science.

As early as 1936, Academician Lysenko indicated in the first discussion that the Morganists-Mendelists have become entangled in the conception of development. And in this lies the fundamental difference between the Michurin concept and the Morgan concept. Michurin's theory is a theory of development. The scientific advance of Michurin lies in that he for the first time showed in succession how development is effected in the independent life of an individual. He showed how in the life of an individual arise and are formed those modifications which later become the basis for the formation of a new species, the basis of phylogenetic development.

Michurin transformed his theory into practice. That is why we say that Michurin raised Darwinism to a higher level.

As regards the Morganists-Mendelists, the theory of development is distasteful to them. At present the Morganists pass over in silence the fact that the very founders of their theory held the viewpoint of the inalterability of the fundamental concept of their theory, the inalterability of the gene, and that they fought long in defense of this inalterability.[1] Even within our recollection, at the time of the first discussion, there were defenders of the invariability of the gene. Now it is inconvenient to point that out, and Comrade Rapoport noted in his remarks that the Morganists acknowledge the variability of the gene. But then variability may differ. It is possible to kill an organism with a club, and this will also be an alteration of the organism, but certainly there will be no development here. The work with mutated gene materials—these are blows of a club on an organism, and therefore the result from them is similar to that of the blows from a club. The Morganists are not in a position to explain how hereditary changes occur. They have dug a deep abyss between modification and mutation. Mutation is not a historical category. It appears at once, like something ready. It is not formed, the environment does not influence its quality. Moreover, it is not connected with preceding mutations. Thus it is clear that being theoretically acceptable in its character of variability, it is impossible to try to influence it. How can a variability, which is not connected with anything, which arises suddenly and ready, be influenced?

That is why the theory of the Morganists does not prepare them for practice but, on the contrary, disarms them.

Comrade Rapoport did not reply directly to the question whether or not he acknowledges the inheritance of acquired characteristics. If he had made an open acknowledgment, he would have said directly that he denies it and in addition denies all the practical achievements of Michurin genetics and all its theoretical aspects.

But Comrade Rapoport could not be so frank. True, it should be said that lately frankness is not characteristic of the Morganists. The frank Morganists are abroad.[2] It is true that one frank expression, obviously at a time when it seemed that the

[1] This falsehood is repeated frequently by the Party liners, and mere repetition, as Goebbels knew, convinces a certain common type of mind. On the other hand, Lewis Carroll satirized such standards in *The Hunting of the Snark*, thus:

"Just the place for a Snark! I have said it twice:
That alone should encourage the crew.
Just the place for a Snark! I have said it thrice:
What I tell you three times is true."

[2] Where they are safe.

Morgan orientation acquired greater strength, made its appearance even in our press. This was B. M. Zavadovski's article "Thomas Hunt Morgan," in which he quite frankly stood up for Weismannist concepts.

Comrade Rapoport and other Mendelists are determined not to admit their origin so openly. The history of science knows such phenomena. Reactionary theories very frequently masquerade themselves, conceal their reactionary nature, conceal their bonds with reactionary theories.

K. A. Timiryazev, who was an irreconcilable foe of such idealistic, vitalistic theories, strikingly characterized their carriers by calling them "forgetters of kinship." Our Soviet Morganists—they are Weismannists who do not remember their affinity. *(Applause)*

Permit me to consider briefly the quality of the evidence which is sometimes brought forth by the Morganists in support of their theory. It is known what great significance is given by the Morganists to the cytological evidence on the accuracy of their theory. One cytological proof, borrowed from plant embryology, is given particular meaning. This is the evidence on the formation of male gametes in angiospermae.

In 1910 one of our greatest scientists, S. G. Navashin, described male gametes, possessing formation of naked nuclei, in a classical object of cytological investigation in the lily Martagon.

In the cytological section of the Grishko and Delon course in genetics, it is stated that this fact has great theoretical significance because it indicates the preëminent importance of the nucleus in hereditary phenomena.

This textbook was published in 1939, the data cited referred back to 1910.

And it is known that Navashin was not only an excellent observer but also a remarkable master of microtechnique. The art of making preparations reached perfection in his hands, and all his life he continued to perfect microscopical technique. He noted the desirability of observation on living material, and with bitterness said that this attempt was not successful.

In this spirit he trained his students. Navashin, as is known, spent the greater part of his creative activity, at the very period when he performed the work which brought him the most meritorious fame, in Kiev, where he established his Kiev school of embryologists. They were trained in the spirit of Navashin to strive for perfection in technique, perfection in observation, perfection in drawing. And it is thus quite natural that over a period of time there began to appear data in the school founded and trained by Navashin which showed that male gametes representing well-formed cells are found in angiospermae. This was first disclosed in the work of V. V. Fink, Navashin's closest student, as well as in the work of other investigators. Thus, in this school was Navashin's will first put into practice—a procedure developed for observation during life with greater magnifications of the microscope. It was shown directly in living material of a large group of objects by Navashin's students—M. V. Chernoyarov and others personally not acquainted with Navashin but trained in his school—that a transfer of the cells in a living, growing pollen tube in which we observe the movement of cytoplasm never produces exposure of the cell nuclei. The sperm always represent entire cells, well-formed cells showing no tendency whatsoever to nakedness.

These works have been only partially published, for example, before the war in the journal *Vernalization.* In this journal are excellently reproduced microphotographs from living material.

These studies were so conclusive that even P. M. Zhukovski noted in his textbook *Botany* that it is necessary to give up previous statements on sperm-naked nuclei, and obviously it is necessary to acknowledge that angiospermae have male gamete-cells.

Everything was excellent. Other studies appeared after the war on living material. Nevertheless, lately obviously in connection with the aggression of the Morganists, sperm-naked nuclei again began to be described. But these studies made use of a faulty procedure, a procedure of simultaneous fixation and staining with acetate of carmine.

Both in our and in the foreign literature it was shown long ago that many delicate formations of the cell are destroyed during preparation by this method. It was again necessary to show that the technique of preparation is significant in the shortcomings of these works. In order to explain this in my latest work, which is being prepared for publication, I worked out a comparative study of living and fixed material. I found so perfect a method of fixation that I obtained cells—sperm—in fixed material of the famous Martagon lily, in which Nevashin did not succeed.

I finally succeeded in establishing that the reason for the descriptions of naked nuclei of sperm was faulty technique. I am taking the liberty of transmitting certain microphotographs and drawings of living and fixed material to the Praesidium. All are of the lily Martagon, a famous subject of cytological investigation.

But it is necessary to mention that the reason for the appearance of the works indicated lies not only in the errors of inexperienced investigators. Ideological deviation is the matter considered here.

Allow me to conclude at this point. *(Applause)*

2

Academician B. M. ZAVADOVSKI

I agree with the entire line of the present attack against the errors of formal genetics, and in doing this, I do not have to be inconsistent. As far back as 1926, in my book *Darwinism and Marxism,* and in all my subsequent statements, I expressed my negative attitude toward Weismannism, Malthusianism, Mendelism, and formal genetics. In 1936, side by side with T. D. Lysenko, I took the field against formal genetics.

I was placed here in the ranks of adherents of formal genetics by a deliberate and unsubstantiated attempt to mislead the Soviet public opinion; I have all the more right to protest, since this was done with the full knowledge of my above-mentioned works and statements. I was placed in these ranks merely because I happen to differ with T. D. Lysenko on some other questions.

Both T. D. Lysenko's report and the statements of his adherents present our Soviet public opinion with a one-sided picture of the condition and distribution of forces in biological science.

Thus, I am of the opinion that T. D. Lysenko is making a big mistake and is misleading our public leadership by letting it appear that in biology there exist but two fronts, or two trends, directed toward the solution of the problems of Darwinism. Actually, both in the historical past, and at present, we can distinguish three trends: one, which I designate as consistent Darwinism, and two others, which are deviations and distortions of Darwinism. One of these deviations consists of the simplified, vulgarly materialistic schemata of mechanistic Lamarckism. The other deviation consists of the reactionary teachings of neo-Darwinism-Weismannism-formal genetics.

Fully supporting the attack developed at this session against the errors of formal genetics, I still believe that T. D. Lysenko ought to have defined more clearly and fully his attitude toward the errors of the mechanist variety. In his report there are indications that, following the spirit and traditions of K. A. Timiryazev, he does not agree either with Weismannism or neo-Lamarckism; it seems to me that these statements in regard to neo-Lamarckism are insufficient and demand more detailed treatment.

One would have liked to hear more clearly about the distinction made by K. A. Timiryazev between Mendelism and "Mendeling." Timiryazev had always stressed that he recognized "Mendelism" as a definite sum of scientific facts and specific regularities which explain actual phenomena of the chromosome-nucleus heredity. At the same time, K. A. Timiryazev fought an impassioned fight against "Mendeling," by which he meant those reactionary, idealistic, and metaphysical deductions and generalizations which our own and the bourgeois Mendelists derived from these valuable scientific facts.

Thus, it would have been more correct to apply the term "Mendelism-Morganism" to that valuable cargo of scientific facts, methods, and investigations which need not be discarded from the ledgers of the Soviet science, since they can be incorporated in their totality into the golden fund of Michurin's trend. Moreover, in their concrete achievements—for instance, in developing the method of polyploidy—these investigations promise to make a material contribution toward an increased productivity of plant cultures. When it comes to practice, no one can say that the methods of polyploidy have failed to acquit themselves well. The varieties of wheat and rye, planted over millions of acres of our Fatherland, have been created by geneticists A. P. Shekhurdin, P. I. Lisitzyn, and P. M. Konstantinov.

A voice from the audience: On what basis?

Academician B. M. Zavadovski: On the basis of many-faceted and many-sided utilization of all the trends and methods created by the Darwinian science.

A voice from the audience: This isn't clear, Comrade Zavadovski.

Academician B. M. Zavadovski: Moreover, this was done not in any contradiction of Mendel's laws, but was often based on them.

Such terms as "Weismannism," "autogenetics," and "formal genetics" would be more applicable to a group of reactionary-idealistic misconceptions current among the bourgeois geneticists; these misconceptions still persist among a considerable number of our own geneticists, in the form of dangerous atavisms of consciousness, and they ought to be eradicated.

It's wrong to criticize our Soviet Mendelists in such a wholesale fashion as we have heard it done here, and not only from the preceding speaker, but from others as well. The level of arguments mainly used here is the level of discussions long past, the level of 1931; I would say that I do not have to answer these arguments, because they are arguments I had used myself, and which I do not repudiate even now. But, comrades, everything develops and grows, and representatives of the Mendelian doctrine in our country rectify their errors to a considerable degree. Moreover, they also contribute valuable achievements to the general fund of our Soviet science and practice.

I do not think that fighting on this particular front is over, but I had a right to expect that the comrades speaking here would show greater discrimination in supporting those Mendelists who are already free of their past errors.

The ideas of Academician I. I. Schmalhausen, as quoted here by Academician T. D. Lysenko, are unquestionably wrong. But I know other works of Academician I. I. Schmalhausen, among them his brilliant investigation of the relationship of conformities of individual and phylogenetic development, which was a serious blow to Weismannism and formal genetics; and knowing this, I would not deem it right to place him in a wholesale fashion in the ranks of formal geneticists, on the basis of some isolated scientific thoughts of his, or even a whole book.

And even if Academician I. I. Schmalhausen is mistaken in some of his statements, you cannot throw into one heap all the many-faceted trends in biology, you cannot heap everything together under the name of formal genetics. That is a falsification. Under the Soviet conditions, there arose new trends, which brilliantly develop Soviet Darwinism on new foundations. And the first place here is occupied by Michurin.

At the same time, I insist that Michurin's trend cannot cover, exhaust, or eliminate all the other trends that exist side by side with the Michurinist trend. First of all, let us take Michurin's trend particularly in the form given it by Comrade Lysenko—the vegetative hybridization of species; surely, no one would seriously offer this in regard to animal organisms. Vegetative hybrids among animals, except for heteropterous butterflies, have not been offered us thus far. Give us concrete directions and proposals how to apply the methods of vegetative hybridization (the prime article of faith of Comrade Lysenko) in regard to the animal kingdom!

Yet such wholesale abuse of any scientist who disagrees with but a single work or statement of Comrade Lysenko sounds very much like suppression of criticism and self-criticism, regardless of personalities.

This is hardly the way for us to form the proper conditions for further creative development of Michurin's doctrine.

In the same order of criticism and self-criticism, I would consider it desirable, for both the speakers and the session as a whole, to give some deeper and more businesslike consideration to those serious divergencies which exist in T. D. Lysenko's views, as compared with Michurin's teachings.

For Michurin, the basic and primary factor in obtaining source material with "unstabilized" heredity was seen in the methods of distant sexual hybridization (geographical or intervarietal). The method of grafting, Michurin regarded, first of all, as a means for the preservation and propagation of the best qualities of the hybrid forms which he selected. And only as a brilliant superstructure of this basic role of grafting did Michurin form his doctrine of interaction between the stock and the graft.

Whereas T. D. Lysenko, both in his previous works and in the report we have heard here, names vegetative hybridization as the primary source for obtaining material with "unstabilized" heredity. In one of his works he even declares that study of the heredity of organisms does not require any sexual crossing.

I do not share in this atmosphere of unconditional agreement and uncritical acceptance of everything said or written by T. D. Lysenko; since I believe that his authority and achievements would only gain in strength, inasmuch as we would help with our criticism and would insist on clarification of all the unclear and inadequately comprehensible aspects of his theoretical views and his experimental investigations.

Those cogitations about assimilation, dissimilation, and metabolism, on which Lysenko bases his ideas of the laws of heredity, are not quite satisfactory. To me, as a physiologist, the idea of F. Engels that metabolism is the basic, the intrinsic character of life, is completely indisputable. But that is precisely why it is not enough to stop at the concept of metabolism, why it is necessary to give it more concreteness; it is necessary to study thoroughly those concrete forms of metabolism, assimilation, and dissimilation, which distinguish such diversified forms of metabolism in the organism as, in one instance, heredity, in another, growth or development, propagation, neuromuscular processes, etc.

Let me now ask: Can we be satisfied with the analysis of the status of biology which was given to us in the report of T. D. Lysenko? No, we cannot. We were told of the widened fighting front and of the task to smash the errors of formal genetics, and that is as it should be. But where is the fighting front against mechanism? Why was nothing said in the report about the task of fighting vulgar simplification in solving the problems of biology and Darwinism? Wasn't it due to the fact that the entire report of T. D. Lysenko was pervaded with an effort to throw into a single heap those who really occupy the formal-genetic positions, and those who are trying to defend the correct positions of the general Darwinian line, and are fighting on two fronts?

And, finally, I cannot help voicing my alarm concerning those contradictions which I had pointed out in my article in the *Literary Gazette,* and which exist between certain statements of T. D. Lysenko on problems of relationship between Darwinism and Malthusianism on one hand, and the general principles of dialectical and historical materialism on the other.

Summing up, I would like to stress once more that it seems to me that all my above doubts and disagreements are fully compatible with the duties and rights of all Soviet citizens in regard to criticism and self-criticism, regardless of personalities. The recent philosophical conference even made it obligatory for us to utilize maximally these basic methods of Soviet democracy. However, not one of my arguments or doubts contradicts in any manner the true task of developing Michurin's doctrine within Soviet Darwinism.

3

TH. A. DVORYANKIN

Editorial Office of Selection and Seed-Culture

Opponents of the Michurinist trend in agrobiology and their allies have appealed to us here to find a rational kernel in the bourgeois biological sciences, to detect "several progressive trends" in biology, and not to lose contact with the so-called world science. It is true that we Michurinists and the Morganists are divided in our different attitude toward both the classical heritage in biological sciences and what the Morganists designate as "the contemporary, world-wide biological science." The content they put into this concept is territorial, and not theoretical. Biological science of the world, at every stage of its development, is represented by the most advanced science. At present the most advanced, materialistic, biological science is the Michurinist trend in the Soviet agrobiology.

Is it not surprising that in our country, where the methodological basis of science consists of materialistic dialectics, whose revolutionary character is unquestionable, there should appear from time to time dreams of unity between our science and the contemporary biological science abroad?

Of course, we Michurinists stand for the classical heritage in science, but we are against uncritical swallowing of all the deductions of the bourgeois professors, even if they deal with classical Darwinism. We remember V. I. Lenin's instruction to utilize all the wealth of knowledge furnished by the developing science, but not to care a straw for the deductions of the bourgeois professors, who see nature through the eyes of men belonging to a bourgeois society.

We take from the classical Darwinism its chief import—the materialistic theory of development of the organic world, as well as the facts that support this theory. But we reject in Darwinism the concept of flat evolution, inseparably bound with Darwin's conclusions, unless they are revised from the standpoint of the Michurin doctrine.

Michurin's key to the transformation of nature corresponds to a higher understanding of nature, a higher level of development of practice and of materialistic biological science, than in Darwin's times. Man must not merely utilize the successive changes in organisms, caused by nature, but he himself must call forth deliberately successive changes that would be of use to him, and then consolidate them and develop them in definite directions by breeding. Michurin's doctrine is the beginning of development of a new Socialist science, free from the errors and limitations of classical Darwinism. Needless to say, there is no comparing Michurin's teaching with Mendelism-Morganism.

Treasure-hunting in selection, agronomy, and husbandry is not for us. We are not satisfied with a mere utilization of natural resources, which is implied in classical Darwinism. That is a philosophy which corresponds with the ideas of the bourgeois class of fortune hunters, who exploit nature in a plundering, wasteful fashion. Our biology can and must transform nature. Not merely to utilize the blessings of nature, but also to increase them steadily. To create new plant forms. To raise the fertility of the soil. This is what is taught by the agrobiology of Michurin and Williams.

Now, when we are fighting for the victory of this teaching, for overcoming the bourgeois influence on biological theory, there arises the so-called "third line in biology." This is how it was christened by Academician B. M. Zavadovski during a debate at the Moscow State University. B. M. Zavadovski said at the time that Academician T. D. Lysenko was wrong in his opinion that there exist but two trends in biological sciences—the Michurinist and the Mendel-Morganist. Actually, it seems, there also exists a third, a true trend—"orthodox Darwinism," whose most distinct expression is presented by Academician I. I. Schmalhausen.

Later, perhaps, B. M. Zavadovski will declare I. I. Schmalhausen to be a "remote past" in science, and will prepare a new line of retreat for Morganism. This new line is already indicated in B. M. Zavadovski's speech at the present session, and is expressed by "the phytohormonal theory of development." The meaning of all these maneuvers is to preserve at all cost the fundamental basis of Weismannism—a denial of the inheritance of characters evoked by the conditions of life.

The tactics of the "third line" in biology are conducted in accordance with this strategic task of "orthodox Darwinism." The tactics of orthodox Weismannists in our country consist in a "Mendelization" of the Michurin doctrine under the pretext that it is necessary to have a "broad" Michurinist trend, as opposed to the "narrow" trend developed by T. D. Lysenko. During the same debate at the Moscow State University, B. M. Zavadovski had advised us "to pay particular attention to the ideological side of T. D. Lysenko's statements," i.e., to search for his ideological errors, so as to render criticism of his works really crushing,[3] and to arouse in our public opinion the necessary sympathy for the Morganists. And finally, to shout about how T. D. Lysenko clamps down on science, and thus to disguise the fact that Morganists are the real bosses in our biological science and scientific institutions.

Here, for instance, is an illustration of the "third line" tactics of B. M. Zavadovski. He spoke today a great deal about "Lysenko's monopolist position," about "law-abidingness," "adulation," and finally went so far as to claim that he is prevented from defending his scientific views. He has complained here that somebody refuses to publish his articles. But let me remind you about such articles as "Darwinism in a Distorting Mirror" by P. M. Zhukovski; "Under the Flag of Innovations" by B. M. Zavadovski; and the articles by Professor A. I. Kuptzov and Professor A. R. Zhebrak in the journal *Soviet Agronomy*.

Adherents of Mendelism-Morganism! You know how to criticize sharply the Michurinist trend, but you've failed to find sufficiently sharp and well-grounded appraisal of the reactionary, bourgeois leaders of biology, who also criticize T. D. Lysenko. Nobody forbids, or can forbid, arguments in science. Let us argue

[3] The really crushing criticism of a scientist in Russia does not consist of showing that he is wrong in his observations or his reasoning, but in showing that his ideology is deviant.

within the single stream of the Michurinist trend about how best to assimilate this teaching, and how to apply it most efficiently to practice. But it is time to demand from you that you too join us in the fight against the reactionary theory of foreign biologists.

The critical current overcomes our Morganists only when it comes to the discoveries made by T. D. Lysenko and other Michurinists. No sooner appeared the theory of phase development and of vernalization, which is based on it, than Dr. I. M. Vasiliev began to argue that the theory itself was probably originated by Klebs; and as to vernalization—it is simply soaking the seeds before planting, a method known since the days of Pliny the Elder.

And what distinguishes the "third line" of B. M. Zavadovski? No sooner were proposed the nested plantings of kok-saghyz, which have considerably increased the yield, than up popped Professor Sabinin and started to prove that this method has been used since the time of Tamerlane. If this is so, then where were you at the time the collective farms were gathering poor crops of kok-saghyz, due to the fact that the agrotechnics of this culture that were offered to them were copied mechanically from the culture of sugar beets?

Here, at this session, the Morganists have assured us of their scientific honesty, repeating it so importunately that their very protestations give rise to doubts. Yet, while falsely accusing Michurinists of distorting the views of Marxist classics on species formation, I. A. Rapoport and B. M. Zavadovski have falsified the ideas of Michurin. They picture him as a repentant Lamarckist, switching over to Morganism, urging us not to exaggerate the importance of rearing methods, and calling upon us to learn from Mendel. Michurinists have more than once exposed the falsity of this Morganist revision of Michurin.

During this entire discussion, the role of the "third line" of B. M. Zavadovski and other orthodox scientists consisted in tying the hands of Michurinists in their criticism of the idealistic theory of "hereditary substance," and in aiding the Morganists against Michurinists. Such is the unsightly role played in science by the "third line" of B. M. Zavadovski and by all those he had lured onto this path. *(Applause)*

4

A. V. MIKHALEVICH

Deputy Editor of Ukrainian Pravda

The Soviet agrobiological science is developing in a close bond with the production of Soviet and collective farms. Every worker in Soviet science must regard as sacred the testament of our great Lenin, who had demanded that with us science should not remain a dead letter or a fashionable slogan, but should become flesh and blood, and become fully and actually transformed into a component of daily life.

This is why I, a Soviet journalist, think it worth while to call the attention of the present session to a series of facts characteristic of the Soviet and collective-farm production in the Ukraine, facts which may bear some weight on the scale of present scientific arguments and discussions of the current problems of Soviet agrobiology.

Let us take, for instance, the fact that cannot but give joy to all the Soviet people—the fact that the Ukraine, having raised excellent crops, is largely successful in fulfilling the granary provision plan. At present, the Ukraine has stored twice as much grain as on the same date last year, and there are quite a few sections which have already completed their primary task; what is more, several sections, only recently ravaged by the enemy, have furnished our State with more grain than they used to before the war. *(Applause)*

Some comrades may say, though it is indisputably a positive fact, what bearing has it on science—the subject of current discussions—and on genetics? I hasten to agree that genetics, formal genetics, and Mendelism-Morganism, whose sterility may be regarded as proved, indeed have no bearing whatsoever on the fact of accelerated fulfillment of the granary provision plan. *(Wild Applause)*

But there is another science—the advanced, Michurinist science, eager and willing to bring all its conquests to the people—science that is developing under the clear, encouraging eyes of Comrade Stalin—science, tied up with the names of Michurin, Williams, Lysenko—and this science, without doubt, has made and is making its contribution to our fight for bread, for an advanced fulfillment of the granary plan.

More than ten years ago, there was a gathering in the Kremlin of our foremost collective-farm (kolkhoz) workers, leaders of higher crop yields, with a view to mutual exchange of experience and of sharing plans for a further struggle for kolkhoz abundance; and it was at that significant gathering that T. D. Lysenko gave a report to our Leader and our best kolkhoz workers about his first successes and daring plans of developing an essentially new Soviet agrobiology. Thousands of kolkhoz workers remember the inspiring comment made by I. V. Stalin during that speech: "Bravo, Comrade Lysenko, bravo!"[4] And since then, the Stakhanovite practices of a number of kolkhoz harvest masters have become even more closely interwoven with scientific creativeness, with the discoveries of Michurinist scientists, with their work toward improvement and transformation of the nature of plants.

"It also happens sometimes that the new trails in science and technics are blazed not by generally-known scientists, but by men completely unknown in the world of science, common men, practicians, factual innovators"—thus spoke Comrade Stalin on May 17, 1938, in his speech at the Kremlin reception for the higher school workers.

At the time, Comrade Stalin had cited, as an example, the Stakhanovites, who surpass all the existing norms established by well-known scientists and technicians. "We see thus what kind of 'miracles' can happen in science," said Comrade Stalin, citing these examples.

If we study attentively our present actuality, we cannot help noticing that this "miraculous" element in our science and in our life has grown within the last few years beyond all measure. The trails, blazed in our science and life by Stakhanovite practices, by the creativeness of masses, become more and more promising and wide.

In this connection, a great deal of attention ought to be paid to the letters to Comrade Stalin, published in our press, which contain obligations undertaken by our various republics, regions, sections, etc. In many cases, these letters deal with obligations of whole regions to achieve a *sharp jump* in the yield of several cultures over a considerable acreage; for instance, when it comes to wheat, to raise the yield over thousands of hectares during one year from 100 to 150 poods [a pood is equivalent to 40 Russian foonts or 36.113 English pounds], etc. This is not merely a fact of patriotism, of heroic labor; such obligations cannot be evaluated except as *scientific daring.*

[4] Whom Stalin hath approved let no man disparage.

For instance, in the Ukraine today, following the initiative of the Cherkassy district (which was considerably aided by the representative of the Lenin Academy, Academician I. D. Kolesnik), dozens of districts are now striving to raise during the current year the average yield of maize up to 40 centners per hectare over the entire planted acreage, i.e., to achieve in a single year a sharp jump in productivity, an increase of 20-25 centners per hectare. Such an increase would do honor to any scientific institution, but one must also take into consideration the gigantic scale of the "experiment" initiated by the kolkhoz masses themselves. This should give 2-3 additional kilograms per work day, and ought to create exceptionally favorable conditions for improving Ukrainian pig-raising; it ought to change in many respects the availability of goods in kolkhozes, ought to raise the material level of kolkhoz workers.

In view of this, the present session ought to evaluate properly the heretofore-unknown-in-science, positive, scientific effect of the Academy on raising the yield of various cultures over large areas. Let me remind you of the gigantic "experiment" with millet in 1939-40, when the Academy, fulfilling the task assigned to it by the Party and Government, furnished wide, scientific, agronomic aid to the kolkhozes and Sovkhozes; together, they had achieved an average millet yield of 15 centners per hectare from 500,000 hectares, and 20 centners per hectare from 200,000 hectares.

V. R. Williams, in the article he wrote shortly before his death, had especially stressed the profound, fundamental significance of this gigantic experiment with millet, as a novel phenomenon in the scientific activity of the Academy, and had insisted on broadening as much as possible such concrete effects of scientific institutions on whole regions of our land.

In the Ukraine, everybody remembers the brilliant results of the aid given by the Lenin Academy, through I. D. Kolesnik, to the Shpoliansk district, where in the droughty year of 1946, they had garnered an excellent millet harvest—27 centners per hectare, all over the planted area of the district. During the past year, 1947, the millet yield of 15 centners per hectare, planned by the Academy Board, was garnered not from just 300,-000 hectares, as given in the assignment, but from an area two and a half times as large!

Of similar fundamental importance are the results with kok-saghyz, achieved by the Kiev kolkhoz workers with the aid of T. D. Lysenko and I. D. Kolesnik.

It seems to me that one of the most important matters of principle, which ought to be noted during the present session, is that, after Timiryazev, Michurin, and Williams, it is impossible to work in biology without taking into account the kolkhoz worker and his active role in transformation of nature.

Remember how Timiryazev used to strive for a contact with the agricultural worker, remember his vegetation booths at the Nizhni Novgorod Fair, etc. Conditions under the czar would not let him become one with the people. But already in the case of Michurin, we see him surrounded not only by fruit, trees, and plants, but always surrounded by kolkhoz workers, horticulturists, teachers, Communist Youth, students, etc. And this is profoundly symbolic. At the present session, as we put on our banner the brilliant principles of Michurin, we must fully evaluate his fundamental, prophetic statements about the new type of agriculturist who, in the person of a kolkhoz worker, has been called upon for historic creation. Michurin based this on the idea that "every kolkhoz worker is an experimenter, and every experimenter is a reformer."

It seems to me the following example will tell a great deal. It is well known that the grass-field system of agriculture, which is the pride and a foremost accomplishment of Soviet science, had and still has, here and there, quite a few enemies, above all in the camp of scientists fawning upon everything that comes from abroad. But let me read to you the opinion of Dimitri Palchenko, a Sitkovetz tractor worker, on the problems of the gramineous system of agriculture. In his letter to the editor, he gives in detail his impressions and conclusions about the Williams' book—a book, by the way, which was recently bought in this Sitkovetz district alone by more than a hundred tractor workers:

> While reading this book, I felt, time and again, as if someone were removing a bandage from my eyes. When I began to apply peeling, and afterwards tilling with a fore-plow, it seemed to me, as if in my brain the science of V. R. Williams has lighted some special beacon lights—strength and knowledge—and they gave me the possibility to see clearly into the very inside of the earth I tilled—this great reservoir of large crops.
>
> I understood then clearly that the structureless condition of the soil, which we see in so many kolkhozes, is a brake on our movement forward. Yet who is going to transform the soils, if not we the tractor workers raised by the Government, by the Party, by Comrade Stalin? . . . And now I see it like this—that the soil is cultivated not only by tractors and agricultural implements, but also by the roots of mixtures of perennial grasses.

The following deduction of tractor-worker Palchenko is characteristic:

> Now that I know the effect produced by grasses, peeling, application of

the fore-plow, etc., I can no longer be indifferent to the way in which grasses are cultivated in a kolkhoz, and whether I shall be sent to till the soil with a plow equipped with a fore-plow, or without the latter. *Should I be sent to a kolkhoz without a fore-plow, I shall buy one with my own money, but I shall till the soil with a fore-plow, and not otherwise. (Violent, prolonged applause)*

Now you just try to push this tractor worker toward the positions of the bourgeois agrochemistry! Why, he has already placed his tractor worker's heavy hand on one side of the scales of the old scientific argument about the gramineous system of agriculture, and his word turns out to be decisive! Probably, this is poorly understood even by some of the scientists, adherents of the system, in Moscow. What is needed is that in them, too, there should burn brighter, as Palchenko puts it, "the beacon-lights" of Williams' science *(Laughter, Applause)* that they too should have more of the practical, aggressive spirit!

Comrades, at this session for several days we have heard more than once a call to truthfulness. Soviet journalists, brought up by *Pravda,* are used to a highly responsible attitude toward this sacred word—truth.

The truth lies in this—it is not just some isolated Michurinists, unacceptable to formal geneticists, but our people, our entirely new working class, our entirely new Soviet peasants, our entirely new intelligentsia, who discard decisively from their science all that is antiquated; all that is against the people; all that is born of servility before the bourgeois West; all that includes survivals of idealism; all that fetters the creative powers of the people and obstructs our forward movement.

The truth lies in this—not only within the walls of the Academy, in institutes, etc., but in life itself, in the practice of millions, the Michurinist trend in biology has been tested, evaluated, confirmed, and accepted with love. *(Prolonged applause)*

The truth lies in this—the most outstanding, profound, and impassioned representative of the Michurin trend, T. D. Lysenko, has also been tested and recognized by the people; he has been tested and evaluated in practice—before the war, when with his works and discoveries he was helping our State to prepare for active defense; in the days of war, when Michurin's science helped to increase the food resources of our Country; and after the war, when T. D. Lysenko is organizing the forces of agricultural science for most active participation in the fight for Communist abundance.

Perhaps some comrades, far removed from daily life, have difficulty in understanding this; but ask some Ukrainian kolkhoz worker who has vernalized grain cultures together with Trofim Denisovich [Lysenko] and according to his advice; ask some Kherson or Odessa kolkhoz worker who saw the increase in cotton yield by his method; ask a Ukrainian woman-kolkhoz worker who remembers well that at a critical moment Lysenko helped to save the beet fields from their terrible pest—the long-nose; ask the rubber growers of Pereyaslav-Khmelnitz and Cherkassy who, with Lysenko's aid, have really achieved miracles on kok-saghyz plantations; ask the kolkhoz workers of the southern steppe regions, where Lysenko's science helped to put an end to potato degeneration; ask a railroad worker in the Urals, who, together with hundreds of thousands of families, had his food situation eased during the war by T. D. Lysenko's proposal to utilize potato tops; ask a soldier who ate the millet which our Country could store up, thanks to Lysenko's jump in productivity in 1939-40; ask all these common people, and they will help you to understand wherein lies the strength of Michurin's science, the strength of the President of the Academy of Agricultural Sciences.

We may say that just as in Maxim Gorky our working class saw itself raised to the very heights of culture, so do millions of kolkhoz workers see in Lysenko themselves, their own creative striving to transform nature, the revelation of their own talents in the fight for Communist abundance. *(Prolonged applause)*

The truth lies in this—that Lysenko has not been alone, for a long time now; that, with the aid of the Party, side by side with him arose a whole new constellation of scientists-Michurinists. They are all men of action, we have heard them here, we know their deeds, they are not bookworms, they are men who recognize their responsibility toward the people. And I think that the truth lies in this—that all the Soviet people are grateful to Comrade Stalin, the Party, and the Government, for their courageous replenishment of the Academy with new Academicians-Michurinists. *(Applause)*

The truth lies, finally, in this—that the triumph at this session of the progressive trend in biology will exercise its effect on the scientific front as a whole, the front so carefully fostered by Comrade Stalin—and it will help the entire scientific front to fulfill its honorable task in the fight to build Communism. *(Prolonged applause)*

5

Docent S. I. ALIKHANYAN

Moscow State University

Academician T. D. Lysenko in his extensive report has raised some extremely pressing problems of modern biological science. One of the basic theses of T. D. Lysenko is criticism of Weismannism.

In the last fifty years, genetics has accumulated enormous factual material. However, more than once, idealists of various shades have attempted to give their own interpretation to the accumulated facts, with the aim of utilizing the data of modern genetics for strengthening their idealistic positions and pseudo-scientific, reactionary conclusions. I have in view the metaphysical, idealistic conceptions of Johannsen, Weismann, De Vries, Bateson, Lotsy, and others. One cannot regard as anything but idealistic the speculative concept of Weismann about the independence of germ pathway from the soma, the body. I have never shared but have always criticized Weismannism, and I consider criticism of Weismannism on the part of T. D. Lysenko as completely right.

Further on, S. I. Alikhanyan gives a critical analysis of various idealistic viewpoints on the problem of the gene of both foreign and our own scientists, such as Johannsen, Bateson, Serebrovski, Philipchenko, and others.

How then can we approach the problem of the gene? A gene is an objectively existing, material particle of the living, material cell. One cannot regard as idealistic the concept of recognition of material foundations of heredity, i.e., genes. Assertion of existence of genes must not be understood in the sense that material particles, present in the chromosomes, i.e., genes, are a substance from which separate characters are constructed. A character is the result of development of the entire cell and of the external environment. The gene determines the specific development of a character. Consequently, in the transmission of a character hereditarily, the decisive part is played by the gene. Whereas in the formation of a character, in its establishment, the decisive part is played by the environment, and in this complex system it is difficult even to say which is the leading factor. When a man controls the development of organisms, then the external environment is the decisive factor.

Talks about how geneticists do not recognize the mutability of the gene and the influence on it of the external environment are entirely without foundation. Environment does act on the gene and changes it.

Our critics assert quite wrongly that genetics ties up heredity exclusively with genes. Such a crude, mechanistic, and metaphysical exposition of the concept of the gene had been developed by quite a few geneticists in the early stages of the development of Mendelism, and has been overcome in the process of development of genetics itself, as a science.

One of the leading modern geneticists, Muller, wrote thus in 1947 about the gene: "Although they are corpuscular in the process of their self-reproduction, their products interact in the cell in a most complex fashion, both with each other and with the products of the surrounding environment, in determining the characters of an organism, contrary to what had been supposed by many of the earlier Mendelians." Thus, the experimental data of modern genetics completely remove the thesis of gene autonomy, of its independence of external environment.

The majority of geneticists work quite successfully in the field of study of factors causing hereditary mutability, and of ascertaining the mechanism and nature of this mutability. The work of our Soviet geneticist, I. A. Rapoport, who has discovered a 100% hereditary mutability under the influence of various chemicals, is the best proof of this assertion. Talks about nonexistence of the gene remind me of the early speeches about the atom. In spite of the fact that nobody has seen the atom in its actual existence, at present already no one doubts the existence of the atom.

After all, there was a time when there were biologists who did not believe in the reality of chromosomes. Yet today we already see under the microscope the real sectors of gene localization and are working on extracting geno-hormones.

Academician Mitin spoke about both the reactionary and the idealistic aspect of the chromosomal theory of heredity. Knowing facts and experiments, one cannot make such statements. After all, here are the words of a man who was formerly one of the sharpest critics of the chromosomal theory of heredity, T. D. Lysenko. Not long ago, in 1947, he wrote: "It is true that certain visible, morphological changes of a given, studied chromosome often, or even always, are followed by changes of certain characters in the organism."[5] Such a formulation will be subscribed to by any geneticist representative of the chromosomal theory of

[5] Can the chromosome theory of heredity be all bad? Lysenko's reply is on page 9.

heredity. Who has discovered and proved this thesis? Our science. Consequently, I ask you: why is it unnecessary and forbidden to study the structure of these chromosomes, seeing that their morphological changes lead to hereditary mutability? It is permissible and obligatory to study them, for we are acquiring a key to ascertaining the mechanism and nature of these changes of the chromosome, and hence arrive at the confirmation of heterogeneity of the chromosome along its entire length, of specificity of different sectors of the chromosome, i.e., in the final account, at the theory of the gene. What's idealistic about it?

Only one thing is clear. One cannot bypass the facts accumulated by genetics, as our critics are doing. It is only necessary to comprehend these facts correctly, from the positions of dialectical materialism, rejecting all sorts of idealistic superstructures over the building of the experimental treasure-house of the chromosomal theory of heredity.

We geneticists have subjected ourselves to little criticism, we did not want to wash our linen in public. This has lead to an accumulation of errors, producing such reactionary concepts as A. S. Serebrovski's teaching of geno-fund, or antievolutionary autogenesis of Philipchenko.

A few words about the laws of segregation, discovered by Mendel. I shall not cite numerous quotations from Timiryazev and Michurin, who had quite distinctly recognized these laws. I only want to point out that Michurin has quite correctly criticized our native "Mendelers," who were expecting to obtain the desired strain in just two generations, by means of utilizing the scheme of segregation. These early Mendelists did not, in all their calculations, take into account the main thing: external environment. It was this aspect that attracted Michurin's attention; he did not completely repudiate Mendel's laws. And that is the essence of the whole thing.

Quite justified is the reproach that we study Michurin and Michurin's heritage insufficiently. We are more occupied with polemics as to who is and who isn't a Michurinist, whereas we still lack a monograph about Michurin. Little attention is given to the theoretical heritage of Michurin. He has turned upside down a whole lot of conceptions in biology, having solved a multitude of scientific problems in an entirely new way. In a series of genetic theses, Ivan Vladimirovich [Michurin] had been far in advance of many geneticists abroad.

I cannot agree with Lysenko, who represents the matter so, as if the fundamental thing with Michurin was vegetative hybridization.

Academician Lysenko: Who has said it and where? Where have I written this?

S. I. Alikhanyan: You are saying all the time that the doctrine of vegetative hybridization composes the fundamental nucleus of Michurin's contents.

Academician Lysenko: You are saying this either intentionally or ignorantly.

S. I. Alikhanyan: I never do anything evil intentionally, I always do everything sincerely and speak the truth.

Academician Lysenko: You, either intentionally or ignorantly, pose the question all the time thus—that Lysenko took from Michurin nothing but vegetative hybridization. I take vegetative hybridization as an educational method. In order that a Mendelist could, at last, understand the role of living conditions in the development and changes of heredity, he must study vegetative hybrids. And so, you will then become convinced that nutrition plays a part. This is what it is all about.

Further on, S. I. Alikhanyan quotes works of A. K. Grell, I. V. Michurin, and continues:

After this, I ask you, Trofim Denisovich [Lysenko], how to tie up with Michurin's views your assertion that by gathering the seeds from either scion or stock, and by planting them, it is possible to obtain offspring of plants, separate representatives of which will possess the properties not only of the strain, from the fruit of which the seeds in question were taken, but also the properties of the other strain, with which the first strain was united by means of a graft. I want to understand this and beg you to explain it to me. It seems to me, that between this statement of yours, and the way Michurin understood this question, there exists a very sharp contradiction. I consider that it was quite correct to speak of the effect of the stock on the scion, in the sense in which Michurin was developing his ideas of the mentor, of vegetative convergence. This is why I think it wrong to say that Michurin repeated Grell's ideas.[6]

In conclusion, I would like only to note that, though I consider quite justified the criticism of idealistic conceptions of formal genetics about the immutability of the gene and of the genotype, about the direct tie between the gene and the character, about absolute immutability and autonomy of the gene, and about underestimation of the role of external factors in the evolution of organic forms; at the same time, we must, on no account, undertake a wholesale rejection of the enormous amount of facts accumulated by modern genetics. This is not in the interest of science development and is harmful to our national economy. I greatly regret that here were not represented extremely interesting works of Sakharov with buckwheat; of M. S. Navashin with

[6] As Lysenko's apotheosis had not yet been announced, the speaker dared to differ with him but not with Michurin.

kok-saghyz; of Khadzhinov with the inbred hybrids of corn; of Glembotzki with caracul sheep; of Khizhnyak with wheat hybrids; of Shekhurdin with wheat. One must particularly note the excellent work of Tzitsin, Pisarev, Zhebrak, and finally, of Astaurov with the silkworm, and many others, who have shown how effectively one can utilize the experimental data of modern genetics and cytogenetics in the agricultural economy.

This is why I consider that it is necessary to utilize all the creative trends of genetic science for the good of our country. Not to lock ourselves within one school, not to be limited by only one kind of possibility, but to involve in science development all the multifarious creative quests.

We reject with indignation the attempts of various English Darlingtons and American Saxes to exploit the creative originality of various schools of our free country in order to create a dissension among us. We Soviet scientists, one of the detachments of the great Soviet people, may argue among ourselves, since in creative arguments the truth is born and science moves forward, but on the base of our moral and political unity. This is what Comrade Stalin teaches us. This is what unites us. This is what is dear to us. And this is the guarantee that we shall give all our strength and knowledge for the good of our Country. *(Applause)*

6

Professor I. M. POLYAKOV

During the last decade, in the capitalist countries there has appeared a multitude of anti-Darwinian evolutionary theories. The bourgeois biologists, unable to solve such important problems of evolution as the interrelationship of external and internal factors in evolution—between organism and environment—or the problem of the part and the whole, etc., have gone over either to the positions of neo-Darwinism-Weismannism (Morgan, Goldschmidt, Shull, Simpson, etc.) or to the positions of Lamarckism (Harms, Weidenreich, etc.). Many biologists who call themselves Darwinists are in reality neo-Darwinists-Weismannists, supporters of autogenetic mysticism.

It was not by accident that many bourgeois formal geneticists arrived at neo-Darwinism. It was not by accident that Morgan, Shull, Hurst, and others stand on the positions of the doctrine of pre-adaptation; that Panker proclaims the stability of Batesonianism; that Trigs revives Lotsianism, etc. These idealistic views are connected with a series of serious theoretical errors in genetics itself (lack of understanding of the physiological nature of the process of hereditary mutability, metaphysical concepts as to structure of the hereditary basis of the organism, etc.). These fallacious views we have encountered also in our country in the works of Philipchenko, Koltsov, Serebrovski, and others.

To us, arguments around these questions appear particularly important, because they are connected with the clarification of the ways to create new breeds and strains, in which our agricultural economy is interested. It is necessary to fight for correct positions in science that would correspond to a materialistic interpretation of life processes.

I suppose that, in the realm of evolutionary theory, we ought to defend and develop Darwinism, freeing it of separate defects and errors. Without Darwinism, without the theory of natural selection, it is impossible to give a scientific, materialistic explanation of organic purposiveness.

The radical fault of Lamarckism does not consist in the fact that the Lamarckists pose a highly important question as to the role of environment in the mutability of organisms, but in this: they start with an idealistic doctrine of primordial purposiveness, and are forced inevitably to complete their views with a recognition of some kind of mystical, externally guiding forces in evolution.

As to the role of environmental factors in changing the hereditary properties of an organism, this role is obvious and great. Hereditary mutability is physiological by nature; it cannot proceed otherwise than through changes of the albuminous [protoplasmic] structures of the organism in metabolic processes. And from this the way becomes clear, in which environmental factors become incorporated into the organism; in which changes, acquired by the organism in the processes of vital activity, can under definite conditions become hereditary. Hence follow the great possibilities in the sense of transforming plant and animal forms in the interests of man, of which unsurpassed examples have been given by the pillar of science, I. V. Michurin.

Michurinist teaching shows us the most fruitful ways of changing and deliberately making over plant and animal forms.

Whereas when we speak of evolution in the wild, natural state we must remember that the variability of the hereditary nature of organisms (of one and the same species) and the variability of conditions of existence of separate organisms inevitably lead to this: the changes that occur turn out to be indefinite in the sense of organism adaptation. "To change does not mean to become adapted," used to say K. A. Timiryazev. In nature there arises a tremendous biological heterogeneity of individuals within the limits of a given species; and the intraspecific contradictions, connected with this heterogeneity, are resolved by selection. This, and not the "Malthusian overpopulation," is the basis of the intraspecific struggle for existence.

The fundamental divergence between Darwinism and Lamarckism is in the treatment of the problem of the creative role of selection and struggle for existence, as a premise of natural selection; the divergence is also in the understanding of the question of the ways of settling organic purposiveness. In this (and not in the question of the role of the environment in mutability) is the root of our disagreements with Academician Lysenko. And if Academician Lysenko has declared in his report that he does not follow along the road of either neo-Darwinism or neo-Lamarckism, then he ought to clarify treatment of the above questions. To hang the labels "Morganism," etc. on Soviet biologists who disagree in anything with Academician Lysenko is not right.

Works of Academician Schmalhausen were subjected here to sharp criticism. In Schmalhausen's works there are questionable theses and errors. To my point of view, however, the fundamental thing is that in his four books Academician Schmalhausen works out materialistically, on a tremendous zoölogical material, a series of most important questions of the evolutionary theory. And it is he who profoundly criticizes autogeneticists, Weismannists, pre-adaptationists, and other varieties of anti-Darwinism. I consider it wrong to bypass it all.

We ought to lead our attack on nature by utilizing all the means and all the possibilities opened by science. The profound and extremely fruitful ideas of K. A. Timiryazev and I. V. Michurin ought to be the general line of development of the Soviet genetics-selectionist science. I, and my co-workers, have been working for a number of years in the field of an important problem of Michurinist genetics—selective fertilization.

Our science, united around the Michurinist progressive ideas, which in the realm of genetics-selection were raised on the shield by Academician Lysenko, will achieve even greater successes.

Academician Lysenko: You speak of the Michurinist trend. Yet in your laudatory review of Schmalhausen's book *Factors of Evolution,* you did not indicate whether you consider it right that Schmalhausen did not mention either Michurin or Timiryazev, not so much as by a single word, not only in the text, but even in the list of literature. Do you consider it right?

I. M. Polyakov: I do not see any reason to oppose the Michurinist trend to all that is positive in Schmalhausen's works. In Schmalhausen's books the achievements of our native science are widely utilized, and in his *Problems of Darwinism,* a number of pages are dedicated to Michurin. A number of important and practically effective methods of changing living nature, which are offered by geneticists working in other trends, cannot be bypassed.

But Darwinism and genetics ought to develop in our country along a wide front.

7

Academician P. M. ZHUKOVSKI

To me, a student of the plant world, the cardinal fact is the alternation of generations in the life cycle of plants accompanied by a change of the nuclear phases. Beginning with the algae, through the mosses and the bracken ferns and the gymnosperms, and ending with the flowering plants, there takes place a single evolutionary rhythm in the change of generations and that is of the nuclear phases. These generations are named asexual or neuter and sexual. The neuter generation is accomplished by the formation of spores, whereby, during this process, there takes place a reducing fission in the nucleus of the mother cell. The diploid phase changes into a haploid one; the spores are always haploid; that is, have half the number of chromosomes. The spore then germinates into a sexual generation, all of whose cells are distinguished by half an accumulation of chromosomes. On this sexual generation are formed male and female sexual cells which, fusing in pairs, restore in the zygote the diploid phase. From the zygote evolves the seed from which springs a neuter generation, all of whose cells are characterized by a double number of chromosomes.

To deny this is simply impossible. There is a change of nuclear phases in the life of the plant; this has been produced by two billion years of evolution and has been verified as a physiological act.

Turning to the animal kingdom, it is seen that there, too, is a reduction of nuclear chromosomes before the formation of sexual cells. Thus, in animal life also there is a single evolutionary line of the haplophase and the diplophase. The bases for it are the single and the double accumulation of chromosomes.

In case there is no change of generations in plants, this phenomenon has also been studied. It is known as apomixis and occurs when the seeds develop without the sexual process. These particular cases lead some people to foresee a further evolution in plants as a liquidation of the sexual process, and thus as an elimination of change in the nuclear phases. I do not think that my opponents in the discussion will agree to such a theory.

Here then it is shown of what importance in the growth of the organic world are the chromosomes which secure the life cycle of the organism. The reducing fission of the nucleus is the most important evolutionary act during which time there occurs not only the most manifold union of parent qualities but also—and this is of special importance—there take place in nature, under the influence of surroundings, mutations in the cells and, above all, in the chromosomes. These changes result in a new genetic effect.

The chromosome theory made it possible for us to ascertain the origin of the cultivated cotton species in which there are distinctly seen accumulations of chromosomes of initial wild growing species. On the same basis this theory elucidated the origin of tobacco, maize, and others.

Lately, the Soviet scientist Jacob Ellenhorn invented a method of differential staining of parent chromosomes; a method based on the difference of electric charges in the father and mother chromosomes. Use of this method makes it possible to see in the split cell the father and mother chromosomes.

Further, Academician Zhukovski speaks about Mendel as an outstanding biologist whom the scientists should worship. He states that Michurin denied the application of Mendelism only to fruit-bearing plants.

I am not an opponent of recasting the nature of plants by breeding; in other words, by methods of action by factors of external surroundings. But to do this is not as simple as my opponent imagines it to be. When I hear that it is possible to remake with two winter sowings hard wheat into soft wheat and that by several similar methods spring wheat can be turned into the most winter-resistant wheat known to us, I doubt it.

If the remaking of the nature of plants is so simple, I ask you to solve practically at least two problems: one in the field of cattle breeding, the other in that of plant culture. Create a mule, that useful animal, with procreative breeding methods. Make the triploid plants, such as apple trees, pears, bananas, and others into normally seeded, normal as regards their utilization for cross-breeding. One often hears complaints that Sakharov's tetraploid buckwheat contains but little quantities of seeds—well, remake it by your methods into a multiseeded buckwheat. If chromosomes are of no importance, prove it on the objects indicated here. Regarding vegetative hybrids, I can only state that up to the present time I have not seen them in a live state. I was promised that they would be shown and demonstrated to me.

Concerning the Michurin doctrine and its revolutionary importance, I can say that I have never and nowhere declared myself against it and that I ardently welcome his teachings. To obtain desired hereditary transmutations is our dream. I, therefore, ask Academician Lysenko to commission the association of his numerous co-workers with the compilation of a detailed guide for obtaining the desired mutations.

I must answer the accusations directed against me as the chairman of the Expert Commission on Biological Sciences at the Supreme Committee that I repressed the Michurin thesis. I admit one error regarding the dissertation of Professor Nuzhdin. We corrected it with his aid and by joint labor in our capacity as fellow members of the Committee. Comrade Nuzhdin is a true Michurin adherent. The theses of such Michurin adherents as Professor Kostryukova, Academician Belenkii, Professor Turbin, and Professor Glushchenko found in me a champion. Some Michurin adherents who are aspirants for candidature have submitted theses which are of a low standard and which hide under the Michurin name. We, of course, have declined them.

In conclusion, I want to say that I crave unity on the Soviet biological front. We who are frequently your opponents are Soviet scientists devoted to our Fatherland, the Party, the Government, and Comrade Stalin, the genius of our great Country. We, of course, do not follow the road of the bourgeois scientists. We aim at the unification of a single Soviet revolutionary biological front. *(Applause)*

8

Professor A. R. ZHEBRAK

The present session shows considerable interest in the factual material on experimental polyploidy and amphidiploidy of cultivated plants. Therefore, I have decided to present, at the current session, a series of factual data on this question. We had begun our work in experimental polyploidy of cultivated plants on the basis of those factual data which exist in modern science.

Modern genetics has established that the number of chromosomes plays a highly important role in the origin of the species of cultivated plants. The genus of wheat is represented by species which, by their chromosomal number, fall into three groups: 14, 28, and 42 chromosomes.

The most widely spread over the terrestrial globe is the species of soft wheat *(Triticum vulgare)*, which is represented by both spring and winter varieties. At that, the most winter-resistant strains exist only among the multichromosomal species. This fact is a convincing proof that the number of chromosomes in cultivated plants has played and is playing an important role in the process of evolution, in the process of adaptation to various unfavorable conditions. The best sort of oats, tobacco, and other cultures also belong to the species with a large number of chromosomes.

The connection between economically useful properties of cultivated plants and the number of chromosomes had attracted the attention of many scientists to the internal structure of the cell nucleus and of its component parts—chromosomes; and also led to elaboration of effective methods with which it would be possible to influence the structure of the cell nucleus, which, according to modern concepts, forms the material basis of the processes of development and heredity. In recent years, geneticists and cytologists have found some highly effective chemical and physical factors which permit us to interfere with the processes of division of the cell nucleus, and to increase the number of chromosomes in cultivated plants.

We had begun our research in obtaining new types of wheat with an increased number of chromosomes in order to create more valuable strains for our agricultural economy. In approaching our research, we knew that a majority of wheat species have originated as a result of an increase in chromosome number, and that further increase of chromosome number will lead to a decrease in fertility; therefore, we have decided to increase the chromosome number not in pure species, but in interspecific hybrids.

Many genera and species of plants, upon crossing, produce totally sterile hybrids of the first generation. Sterility of hybrids is due to dissimilarity of nuclear structures, dissimilarity of chromosomal complexes. Yet if these sterile hybrids have their chromosomal number doubled, their fertility is restored, and entirely new types of plants are produced. Such new types are called allopolyploids or amphidiploids. We have obtained amphidiploids with nine different wheat species.

In the last few years, we have succeeded in obtaining, by

chromosomal number, three groups of wheats with 42, 56, and 70 chromosomes. One of these groups, the 42-chromosome one, corresponds with wild soft wheat and permits us to reveal the process of origin of natural 42-chromosome species.

In nature, soft wheat originated not from hard wheat, but from hybrids of hetero-chromosomal species, with the subsequent doubling of the chromosomal number. We have produced experimentally the 42-chromosome wheat by crossing *Triticum durum* with *Triticum monococcum* (14 chromosomes). First-generation hybrids of such crossings have 21 chromosomes and are sterile, whereas upon doubling this number of chromosomes, we obtain 42-chromosome types which become normally fertile and do not segregate into original species. Most likely in natural conditions too it is precisely in this fashion that the 42-chromosome species of wheat had originated in the past.

The 56- and 70-chromosome wheats were first obtained in our Country; they did not exist before in natural conditions. We have succeeded in obtaining new types of wheat that nature itself has not created heretofore. These new types of wheat represent new, valuable material for further practical selection, and as material for studying the problems of species-formation. In particular, the most valuable results were obtained by us upon further hybridization of these new 56-chromosome and 70-chromosome types of wheat with the soft wheat *(Triticum vulgare)*.

Upon crossing 56-chromosome types with soft wheat, the 42-chromosome kind *(Triticum vulgare)*, we obtain in first-generation 49-chromosome hybrids, which are of low fertility (they begin to develop isolated seeds in the wheatear). However, in subsequent generations, as more balanced types are obtained, fertility increases, and in following generations there is manifested a tremendous process of form creation. In the fourth and fifth generations, we have succeeded in obtaining forms with very high fertility, so far as spikelet, spike, and plant are concerned. Some plants had over 1,500 grains, over 100 grains per spike and 7-9 grains per spikelet, with the absolute weight of 1,000 grain being 40-50 grams, and hyaline grain. Among the families of the fifth generation, was obtained a series of lines which completely coincide with the species of wheat existing in nature, such as *Triticum compactum, Triticum vavilovi.*

Academician Lysenko: What you are telling is a commonplace thing. Go, visit Gorki, and you will see there everything.

A. R. Zhebrak: Among the same generations, there are also types similar to genera that are close to cultivated wheat. The facts of experimental polyploidy in wild and cultivated plants presented here are a substantial proof of the correctness of the modern chromosomal theory of heredity.

Besides working with the method of distant hybridization and experimental polyploidy, we conduct work in selection of wheat also by the customary methods of modern genetics—the method of pure lines. By means of this method, we have reared four valuable lines, two of which, in 1946, were turned over to the State for strain testing. According to their inspection in the Moscow region, they yielded in the Dmitrov sector 41.9 centners per hectare and 41.1 centners per hectare respectively, whereas the standard yield there is 39.2 centners. In the Podolsk assortment sector, they gave correspondingly 31.8 and 29.9 centners respectively, while the standard there yields 29.8. Thus, in these assortment sectors, lines of our selection surpass the standard. One of our lines does not fall at all, and represents great value for harvesting with combines. Our hybrids of polyploid wheats of the fourth and fifth generations are characterized by even higher indexes than the best of our lines raised by the usual methods of modern genetics.

Besides wheat, our laboratory procured tetraploid types of buckwheat and millet in several strains. Autotetraploids in these cultures are less fertile than the diploids; however, the absolute grain weight increases 25-30%. To increase the fertility of the new tetraploid types of buckwheat and millet, we have applied the Michurinist method of crossing strains of different geographic origin. Analogously with the laws we have discovered for wheat, the fertility of hybrids of the tetraploid strains will be increased.

Works in producing basically new types of wheat, buckwheat, and millet by means of the action of chemical elements on the nuclear structure of the plant cell show that it is precisely the nuclear structure which is the material basis of the processes of heredity and development, and that it depends on the factors of the external environment.

Combining hybridization with the effect of strongly acting chemical and physical factors, it is possible, experimentally, to create new plant species. Recent experimental works with distant hybridization and with the effects of colchicine and apenaphtene and other factors show that modern genetics has mastered the process of evolution and is able not merely to explain the evolutionary process, but even to create new plant species.

Works on experimental polyploidy of cultivated plants are a convincing proof of the correctness of the chromosomal theory of heredity. Further progress in the domain of genetics and selection of cultivated plants lies along the path of mastering the laws of all the processes occurring in the cell, and of studying how they are affected by the external environment.

VIII

The Climax of the Arguments

THE MORNING SESSION OF AUGUST 6

There was no afternoon session on this day but a total of ten papers was presented in the morning. Papers 1, 3, 8, and 9 were published in *Pravda* on August 8. Papers 2, 4, and 10 were published on August 11. Papers 5 and 7 appeared on August 9 and paper 6 on August 10.

This was the last session for the presenting of papers by the participants of the conference. Several of the best orators were reserved for the occasion and the meeting successfully avoided anything anticlimactic. All of the papers are either interesting or important, or contain at least some interesting passages. Two of the geneticists who had been attacked in previous sessions were given a chance to defend themselves. The other participants, the Lysenkoists, abandoned all pretense of scientific calm. On two occasions the witch-hunters came out into the open. Even the two scientific illiterates are worth reading because of their stupid boners.

The session was opened by I. I. Schmalhausen (§ 1), who had been attacked frequently by the earlier speakers. He defended himself, but did not defend scientific genetics. Like Zavadovski (Ch. VII, § 2), he felt apparently that it was impossible to rescue what had been smeared so thoroughly in the presence of the group, which had itself done the smearing.

The second speaker, I. N. Simonov (§ 2), urged the purging of all geneticists from the Timiryazev Academy. S. Th. Demidov (§ 3) attacked the geneticists personally and started the hunt for heretics and heresies. D. A. Kislovski (§ 4) established his own orthodoxy and then proceeded to criticize Lysenko on a very minor matter. He suffered no ill effects as he demonstrated Lysenko's ability to take criticism and thus helped to refute the charge that Lysenko would not tolerate any opposition. I. Th. Vasilenko, one of the illiterates (§ 5), showed what he believed about the agriculture of the United States and about the milking machines used in the U. S. A. A. N. Kostyakov (§ 6), another illiterate, claimed that a fact which was really known to neolithic agriculture was a discovery of V. R. Williams. P. P. Lobanov (§ 7) in a speech frequently interrupted by applause put biological research on a proper Marxian basis.

V. S. Nemchinov (§ 8) defended his administration of the Timiryazev Academy very ably. His talk was punctuated by hostile remarks from the audience. V. N. Stoletov (§ 9) continued the attacks on the geneticists on both a personal and a doctrinal level. His speech led up to the grand climax, an oration by I. I. Prezent (§ 10).

Prezent's paper should be read by all interested in the freedom of science. It is particularly important that it be read by those scientists who have succeeded in remaining ignorant of the treatment afforded scientists in Russia. Prezent, like Mitin, is intelligent enough to know just what he is doing, but he made a number of careless slips which are incidentally very valuable and very educational for American geneticists.

1

Academician I. I. SCHMALHAUSEN

In appearing here, I want to reply to the accusations that were advanced against me.

The first accusation is one of autogenesis. I have always tried to stand on the positions of a materialistic explanation of evolution, and have consistently fought against idealism of any shade whatsoever. Here, an attempt was made to place me in the camp of geneticists, and at that, formal geneticists. For those who are not well informed on the subject, I must say that I am, in general, not a geneticist, but a morphologist, an embryologist, a phylogeneticist. Only my work on phenogenetics of the racial characters in hens has some bearing on genetics. My works bear no other relation and did not bear any other relation to genetics; all the more they bear no relation to formal genetics.

I had tried to be a consistent materialist, and it seems to me that this is expressed sufficiently clearly in all my works. It is precisely from these positions that I had criticized all those idealistic views which have been ascribed to me here. In *The Problems of Darwinism,* you will find criticism of Weismannism, and of De Vries, and of formal genetics, and of the views of Lotsy, and of the theory of pre-adaptation.[1] Many of these theories—for instance, the theory of pre-adaptation—were first subjected to such a thoroughgoing criticism in the Soviet Union precisely by me.

On what, then, rest these accusations of autogenesis and, therefore, of idealism?

Apparently, they concern the question of the sources of mutability.

I consider that the source of mutability lies in the external environment, but that this mutability is realized in an interaction between organism and environment, and that the specificity of mutation is determined more by the organism than by environment, in view of the greater structural complexity of the organism.

I am being blamed for stressing the indefiniteness of mutability of organism. But I am speaking of indefiniteness of only new changes, and not of indefiniteness of reactions in general.

In the process of evolution, under the creative influence of natural selection, they are transformed into adaptive changes.

These changes, as a rule, are hereditary. It is clear that, basically, they are those changes which we now call mutations. Yet partially they also include their combinations, and the related, nonadaptive modifications. In natural material, we are, in practice, always dealing with indefinite individual distinctions.

There was ascribed to me a conception of evolution as proceeding along a declining curve, corresponding to the concepts of Daniel Rose and other bourgeois theoreticians.

It seems to me that I was the first among the Darwinists to note precisely an acceleration of the process of evolution, rather than its extinction.

Here was also mentioned the theory of stabilizing selection, yet nothing was said directly on the subject. Analysis was made

[1] Pre-adaptation, unfortunately, is one of the more complex factors in evolution. Faced with explaining it to Michurinists, almost anyone might be excused for simply dismissing the subject. We may describe it as follows. Adaptation implies a certain amount of adaptability. This is because most environments vary from time to time, and tend to fluctuate around a norm. Thus, to be successful, a plant must be able to live through both wet and dry years, hot and cold periods, etc. This adaptability may be achieved by morphological changes and a given genotype may appear different in different environments, e.g., a shade leaf may differ from a sun leaf, etc. When a plant is removed from one environment to another which is different, but still within the range of the plant's tolerance, the plant should change adaptively. In the new environment, selection also works differently and in time the plants in the two environments should acquire a different gene frequency, and thus diverge in their evolution. Pre-adaptation as an evolutionary factor gives organisms a certain amount of flexibility and enables them to weather a change which otherwise might cause them to become extinct. Among the biologically untrained, the effects of pre-adaptation have uniformly been interpreted to be the result of the inheritance of acquired characters.

of only the concept of the "reserve of hereditary mutability," which has no necessary connection with the theory. This concept was introduced by me in counterdistinction to the concept of "genic fund" of the geneticists, to designate the entire mobile reserve of hereditary changes of entire populations.

In this "reserve," of course, there is no store of adaptive mutations, ready for all occasions, which Academician Prezent ascribes to me.

Never, nowhere, did I, or could I, speak of anything like this. On the contrary, I had constantly argued with the geneticists on this question, since I regard all mutations as harmful. This means that I could not have spoken of adaptive mutations and their accumulation in the reserve.

I had introduced the concept of a reserve of hereditary changes precisely in order to counterbalance the concept of formal geneticists concerning the fund of genes. If the genic fund is a static concept, then the reserve is a dynamic one. In the reserve there occurs not only the spending of the hereditary material, but there takes place a continuous accumulation of hereditary changes. In *The Factors of Evolution,* I speak of accumulation of the reserve at the expense of mutations, the spread of mutations, and their combination and transformation into complex hereditary changes. Particularly freely does accumulation of hereditary changes proceed under the conditions of domestication; it means that in domesticated animals and cultivated plants we have a maximal accumulation of hereditary changes.

I have never given a positive evaluation of separate mutations. A special book has been devoted by me to the question of evolution of the organism as a whole. In it I prove that only changes of the entire organization as a whole could be useful to the possessor of these changes. Separate, partial changes cannot be useful. This is why any mutation is harmful. Never and nowhere did I speak about utilizing separate mutations, and all the more I did not recommend it to selectionists. I always spoke about complex hereditary and nonhereditary changes.

Finally, the last important accusation is the accusation of disarming the practice. Already from the above, it may be seen that this accusation is groundless. The following statement, as though made by me, is attributed to me: that "the formation of breeds, stormy at the dawn of cultures, becomes gradually extinguished." Not in one of my books, not in one of my articles, not in one of my reports is there such a statement. I have never asserted this.

Remember the selection of sugar beets for sugariness, which began only in the last century, and not at the dawn of culture. It has led very quickly to a limit. To fight the limitation, if only it is not a physiological limitation, is possible, and I definitely indicate the means of increasing the mutability—hybridization and action of external factors.

The last accusation is: why don't I speak about Michurin, why don't I speak about other achievements of our selectionists? Because the book *Factors of Evolutions* is not devoted to these problems.

In connection with this, I have there almost no references to any works prior to 1920. And I have no references to the classic authors. I take the material which I needed for the basis of the theory of stabilizing selection, and nothing more.

Simultaneously with this volume was published my other book, *The Problems of Darwinism*. Is it possible that I was obliged to repeat in one what was written in the other? In *Problems of Darwinism* I present both the history of the problem of factors of evolution and an exposition of the practical conclusions of Darwinism. In this book, a great deal of attention is given to the classics of Darwinism, and particularly to K. Timiryazev, and also to the really remarkable achievements of Michurin. There I also devote enough space to the accomplishments of Academician Lysenko, Academician Tzitsin, and other Soviet selectionists.

It seems to me I have noted all the basic reproaches addressed to me. Perhaps, though, only one more remark.

Prezent accused me of attributing too great an importance to hybridization. It would seem strange, for Michurin too attributed great importance to hybridization. But, it turns out as if I make it an absolute requirement for the existence of evolution.

I am not so illiterate as not to know that bacteria, for instance, lack sexual process, and yet evolve. I do note the tremendous importance of sexual process and crossing in the evolution of organisms, particularly the higher ones. But it does not mean, of course, that it is a necessary condition of any evolution. Hereditary changes appear independently of hybridization. And yet, meantime, Prezent deduces that I reputedly attribute the same importance to hybridization as did Weismann in his theory of amphimixis. Weismann's theory of amphimixis supposed that crossing is the source of mutability. I categorically deny this. I speak of the external factors as sources of mutability; whereas as to crossing, I regard it as a means which permits a quicker combination and synthesis of expression of separate mutations. It is clear that my concepts have nothing in common with the long-ago repudiated theory of amphimixis.

At this point, allow me to end my explanations concerning the remarks made here about me.

2

I. N. SIMONOV

Candidate of Agricultural Sciences

The history of K. A. Timiryazev Moscow Academy of Agriculture is essentially the history of the great division in the biological science of our country. Here for many years labored the leaders of biological science, K. A. Timiryazev and V. R. Williams. Here, for the first time, even prior to the Great October Revolution, V. R. Williams laid the groundwork of selection of agricultural plants.

But does the Timiryazev Academy serve today as a center of application of Timiryazev and Michurin trends? In answer to this question, let me state that today this Academy does not represent such a center. To show that these are not unfounded statements, let me cite some examples.

Everyone at the Timiryazev Academy who adhered to the views of Michurin and Lysenko was under various pretexts "promoted" and transferred to other institutions, or else simply dismissed. This is what was done with the Scientific Collaborator, and now Academician, Ushakova; with the Scientific Collaborator Tikhonenko; with Professor Veprikov; with the Michurinist-horticulturist, Pavlova; and with many others.

Things reached such a state that the Director of Timiryazev Academy, Academician V. S. Nemchinov, started simply to banish from the Academy all the teachers he did not like, teachers who, at the time they joined the faculty, had been recommended by I. V. Michurin. And all this was done merely because at some time, some place, these comrades had been favorably mentioned either by I. V. Michurin himself or by his pupils.

Within the walls of the Academy, many still maintain a slighting attitude toward the teaching of I. V. Michurin. Here, they have forgotten the letter written by Michurin to a students' circle, and published in the first volume of his *Collected Works* (page 244), in which he stressed that ". . . due to complete lack of information on the subject, the attitude toward him was much too slighting, and in some cases downright ignorant."[2] That's how it was at the Academy at the time Michurin wrote these lines. And that's how it remains even now.

Academician I. V. Yakushkin, in his speech here, has noted

[2] Did Michurin suffer from paranoia?

quite correctly that adherents of the non-Michurian trend are growing fewer and fewer. However, in regard to the scientific body of the Timiryazev Academy, such a statement would not be completely accurate. Nor can one apply it to Academician V. S. Nemchinov himself, the Head of the Academy. Only the following at the Academy may be regarded as belonging to the Michurinist trend: Academician I. V. Yakushkin, Academician V. P. Bushinski, Professor V. I. Edelstein, Professor N. N. Timofeev, Professor M. G. Chizhevski, and their co-workers.

I may boldly assert that the Director of the Academy, Academician V. S. Nemchinov, himself holds the views of Mendelism-Morganism.

Whence, from what universities, from what professorial chairs, were sent to America, for its crafty, reactionary press, articles containing crude attacks on Academician T. D. Lysenko? Why, they came from no other place but from the Timiryazev Academy. They were written by Professor A. R. Zhebrak, who spoke at this session. They were encouraged by the support on the part of Academician V. S. Nemchinov.

These scientists, such as A. R. Zhebrak and others, apparently even now—and I am fully justified in saying this, judging by their speeches here—are not free from worshiping reactionary science. *(Noise in the audience)*

Let us recollect, comrades, the Session of Horticulturists held two years ago, and the obstruction created at the time by the students of the Timiryazev Academy against Michurin's best pupil, Academician P. N. Yakovlev.

Two years ago, the Timiryazev Academy had announced a competition for the supervision of the Chair of Technology of Fruits and Vegetables. Among the competitors was a student of Academician A. N. Bach, a biochemist, and now a Stalin Prize laureate, Professor B. A. Rubin. But he is an adherent of the Michurin doctrine, and therefore he did not get this post.

The Academy does not conduct any creative, selectionist work. And this suits Academician V. S. Nemchinov fine.

And here is another example. Five years ago a small group of workers of the Michurin Fruit and Vegetable Institute, of which I was then in charge, had made up a new program of selection and genetics. The Institute realized even then that the program of selection and genetics for agricultural colleges was antiquated and demanded a radical revision in the spirit of teachings of Michurin and Timiryazev. This program was sent to V. S. Nemchinov for his opinion, but the workers of the Michurin Institute failed to receive any reply.

The Timiryazev Academy must reconstruct its work in a thoroughgoing fashion. Its students must study thoroughly the works of Michurin, Timiryazev, Williams.

Some speakers here tried to accuse Academician T. D. Lysenko, implying that he stood for nothing except vegetative hybridization and a one-sided assimilation of the Michurin heritage. This is slander against Academician T. D. Lysenko. It is enough to examine the ten-page preface to the first volume of Michurin's works, written by Academician T. D. Lysenko, to see that at that time he had already indicated the tasks of assimilating Michurin's doctrine.

Academician T. D. Lysenko always stands for a new, progressive teaching, and we know that new, progressive forces win because they are called to life by the conditions of development, by reality itself.

And it is totally unbecoming to some of our scientists to resurrect the nonexisting laws of bourgeois science—the laws of Mendel and Morgan—and to forget the mighty heritage of Michurin, Timiryazev, and Williams.

Everyone of us is certain that the names of our outstanding Russian scientists—Timiryazev, Michurin, Williams—shall live forever in the hearts of our many-nationed Soviet people and in the hearts of all progressive mankind. *(Applause)*

3

Academician S. TH. DEMIDOV

The present session of the Lenin All-Union Academy of Agricultural Sciences will undoubtedly play a highly important role in the further development of the agronomical science in our Country. We are dealing here with a matter of highly important principles, important not only for such branches of science as genetics and selection in plant-breeding, the doctrine of livestock-breeding, and problems of pedigree breeds in animal husbandry. The report by Academician T. D. Lysenko and the questions raised by the members of this session during the debates are directly connected with the daily struggle of the collective and Soviet farms for a successful solution of the chief problem of the postwar, Stalin five-year-plan—increase in the crops of farm cultures, in the total production yield, in the number of livestock, and in the latter's productiveness.

The report by Academician T. D. Lysenko and its expanded discussion at this session have shown quite definitely that in biology there have taken shape two trends, distinct and opposite in principle as to their ideological sources. One of them—an advanced and truly scientific trend—is the Michurinist trend; the other is the reactionary, Mendelist-Morganist trend. All the attempts to reconcile these conflicting trends in biology, to take some sort of an intermediate position between the Michurinist and the Weismannist (Mendelist-Morganist) trends—as it was advocated at this session by Academician B. M. Zavadovski and Academician P. M. Zhukovski—are doomed to failure.

The existing contradictions between the two trends must be resolved not by way of a reconciliation but, in principle, by a sharp, open struggle.

Michurin's teaching in biology helps the selectionists and practitioners of the Soviet farm economy to create new forms of cultivated plants and new breeds of farm animals.

Michurin's trend in biology proceeds from the premise that changes in the conditions of life, conditions of the external environment, will inevitably lead to this: these new conditions of life, sooner or later, but without fail, will break up the old type of development of plant and animal forms and will create new structures of these forms, *corresponding* to the new conditions of life. Michurinist genetics has proved in practice that the new properties of plants and animals, acquired by them under the influence of the changed conditions of life, can be transmitted hereditarily. In this lies the great progressive significance of Michurin's works. It is precisely because of this that we are fully justified in saying that Michurin's teaching represents a fundamentally new stage in the development of the evolutionary theory and of biological science. Michurin had advocated an active interference in the workings of nature, and thus had opened unheard-of perspectives for an accelerated formation of new forms and strains of cultivated plants, of new animal breeds, and of guiding their development in a direction most useful for man. For us, as Michurin used to say, "at the moment, the most essential problem is to find a way, to find a method which, upon being fully grasped, would make it easier for us to interfere more successfully with the workings of nature, and in the very process of doing it, to disclose its secrets."

Advanced agronomical science is characterized by the fact that it arms practitioners for a further uplift of the Socialist agricultural economy. It is precisely this kind of science that we see in the Soviet agrobiological science, created by Williams and Michurin, and being successfully developed by Academician

Lysenko. One of the greatest services rendered by Academician Lysenko consists in this: on the basis of solving practically important problems of development of Socialist agriculture, he has skillfully combined the teaching of Timiryazev-Michurin about form-creation and transformation of the nature of plants, with the teaching of Dokuchayev-Kostychev-Williams about soil formation and methods of increasing soil fertility, the teaching of grass-field system of agriculture.

Soviet agrobiological science, created by the labors of Williams-Michurin-Lysenko, is highly fruitful. This science has worked out such radical measures for raising crop yields as the correct gramineous and fodder crop-rotations; application of the scientifically grounded system of the basic and preplanting tillage of soil—fallows and plowings; assimilation of the system of organic and mineral fertilizers; and determination of the place fertilizers have in crop-rotation. This science has created an orderly system of measures for improving seed-cultures on collective farms and Soviet farms; has established scientific foundations of development of irrigation agriculture and methods of fighting the processes of secondary soil salinity; and has worked out the problems of creating a system of agricultural machinery adaptable to the peculiarities of different agricultural zones, a system of machinery corresponding to the conditions of application of advanced agrotechnics. All these measures are becoming more and more incorporated in the production of collective and Soviet farms and permit our country, year after year, to enlarge the crop yields and to increase the productivity of animal husbandry.

I consider it necessary to note the vastness of the scale on which the proposals of the representatives of Michurin's trend, above all of Academician T. D. Lysenko, have become incorporated in the Socialist agricultural production. Such a generally known agronomical method as *vernalization of seed-cultures*—which allows valuable strains of summer wheat to advance into more northerly regions, and which insures a considerable crops' increase—is an obligatory condition of the State plan of agrotechnical measures on the collective and Soviet farms. In 1940 plantings with vernalized seeds were conducted over an area of 13 million hectares. The State plan of agricultural development for 1948, in accordance with the decisions of the February plenum of the Central Committee of the Communist Party of the Soviet Union, provides for planting seed-cultures with vernalized seeds over an area of 7 million hectares.[3] The planting

[3] Quite a reduction in acreage. Were these seeds actually vernalized or were they only descended from seeds which had been vernalized?

of vernalized potato tubers is being conducted over considerable areas. *Summer plantings of potatoes* guarantee to put a stop to degeneration of the potato-planting material in the droughty steppes region of the south. The method of summer planting increases the yield and considerably improves the strain qualities of potatoes. A strain of winter wheat, "Odessa-3," established by the Odessa Institute of Selection and Genetics, under the guidance of Academician T. D. Lysenko, surpasses the yield of standard strains by 3-4 centners per hectare, and is both frost and drought resistant. The strain of cotton "Odessa-1" is, in essence, the basic strain for the new regions of cotton culture.

It is particularly necessary to note the wide, productive assimilation of measures *to increase the yield of millet.* The Lenin All-Union Academy of Agricultural Sciences, under the direct leadership of Academician T. D. Lysenko, had, as far back as 1940, offered a thoroughly thought-out system of measures as to dates and methods of planting, cultivating, and harvesting millet. The realization of these measures, over an area of more than 500,000 hectares in 1940, and more than 1 million hectares in 1947, has guaranteed millet yield of 15 centners per hectare and more. For several years, on 85-90% of the area of all cotton cultures, the method of cotton stamping has been used, which insures its preservation from shedding ovaries, and increases the prefrost harvesting of the best strains of raw cotton 10-20%.

During the years of the Great Patriotic War, Academician Lysenko has made proposals to insure higher germination of seed-cultures in the eastern regions of the USSR, by means of warming them. Incorporation of these proposals has permitted the Siberian collective and Soviet farms to increase considerably their own seed resources, and to increase crop yields. Methods, worked out by Academician Lysenko, for installing winter-wheat cultures in the regions of the Siberian steppes have, at present, been tested in practice and are supported by the agronomic public opinion and by the local party and Soviet organs, especially by the Altai Regional Committee of the Communist Party and the Regional Executive Committee. *(Applause)*

Representatives of the Michurinist trend in biological science have worked out, and have widely applied in practice, such effective methods of selectionist work as intrastrain and interstrain crossings, methods of rejection in the process of selection, and a deliberate selection of parental pairs.

In accordance with the decisions of the February plenum of the Central Committee of the Communist Party of the Soviet Union, in the steppe regions of the south, at present, summer plantings of lucerne on fallows are being widely incorporated;

this quickly insures a considerable increase in the yield of the seeds of this culture, which is so necessary for the assimilation of proper, gramineous crop-rotations.

These and other achievements of the representatives of the Michurinist trend speak convincingly for the fact, that one of the basic singularities of their scientific work is their everyday tie-up with the Soviet and collective farms; their drawing into scientific research of a large contingent of the vanguard workers of agriculture; and a rapid incorporation of scientific achievements into agricultural production.

Experience of thousands of advanced collective and Soviet farms shows graphically that it is precisely assimilation of the system of measures, worked out by Williams-Michurin-Lysenko, which is one of the most important requisites of their production successes. This is the reason why our State plans for the development of the Socialist agricultural economy include, without fail, the tasks of incorporating the achievements of the advanced agronomical science. Every year our Government approves the plans for raising cultivation of black fallows; for tilling with a plow equipped with a fore-plow; for husking the stubble; for planting seed-cultures with vernalized seeds; for conducting summer plantings of potatoes in the south; for adopting sound gramineous crop-rotations; for planting field-protecting forest zones; for developing irrigation, based on utilization of water of the local watershed; for wider plantings of high-yield strains, established by the Michurinists.

At the same time, opponents of the progressive Michurinist teaching—representatives of the reactionary-idealistic trend in biological science—Weismannists (Mendelist-Morganists) gave nothing of value in the development of the Socialist agricultural economy. Representatives of formal genetics, such as Professor A. R. Zhebrak, Professor M. S. Navashin, Professor N. P. Dubinin, Docent S. I. Alikhanyan, and others, limit themselves to sterile, cabinet experiments with a fruit fly, and to raising tetraploids and polyploids. What could the Weismannists (Mendelist-Morganists) show us, what could they tell us from the rostrum of the current session? Absolutely nothing! Except, perhaps, for some tetraploid grains of buckwheat in test tubes, or else that fly from Voronezh, studied by Professor Dubinin, in which, as a result of the war, there occurred a transformation of the chromosomal apparatus. *(Laughter in the audience)*

Loss of contact with life, sterility, and the reactionary essence of the Weismannists (Mendelist-Morganists) may be strikingly illustrated by the contents of the latest book by Academician I. I. Schmalhausen, *Factors of Evolution.*

In this work Academician I. I. Schmalhausen regards formation of breeds in farm animals and formation of strains in cultivated plants, as a ferreting out of some hidden, reserve mutations which had taken place prior to cultivation, or else in the incipient stage of cultivating the given species of plant or animal. These reserve mutations, or "reserve of mutability" as the author calls them, were, according to Schmalhausen, merely revealed in the historical process of breed or sort formation, and not created by culture. On this basis, the author asserts that the most intensive formation of breeds and strains occurred at the dawn of culture, while "subsequent, directed selection proceeds already slower. . . ." This is one of the basic theses of Schmalhausen's book, and it contradicts the facts and disarms practitioners-selectionists in their work of raising new strains of agricultural plants and new breeds of animals, since it leads to an assumption that the "golden age" of selection is already past. From the point of view of this theory, Academician I. I. Schmalhausen would be powerless to explain the causes of the phenomenon he describes in his book—the abundant species-formation among pigeons, and the small number of varieties among such domestic birds as geese and ducks. According to Darwin, the small number of varieties of geese and ducks is the result of one-sidedness of man's interests in establishing the races of these birds (varieties of geese and ducks were created for the sole purpose of obtaining meat, eggs, and feathers) ; while the abundant species-formation among pigeons is the result of diversity of man's selectionist aims in rearing pigeon races (various economic purposes, sports, military, mail, decorative, and other interests).

Whereas, according to Schmalhausen, the small number of varieties of geese and ducks is to be explained by their having a "smaller reserve of mutability." It follows from this that pigeons represent the most progressive branch of the animal kingdom, while geese and ducks are nonprogressive, since they do not possess an adequate "reserve of mutability." Such are the reactionary and nonsensical conclusions one can reach as a result of profoundly erroneous, fundamental, theoretical positions in biological science.

Academician Schmalhausen's book advertises in every way the reactionary-idealistic teaching of Mendel-Morgan, and of other reactionary authors abroad, for it does not mention by as much as a single word the name of Michurin, and Timiryazev is ignored completely, in spite of the fact that both these leaders of Russian science have devoted their whole long lives to the study of the problems of evolution. Such facts are unworthy of a Soviet scientist.

The fact that Academician I. I. Schmalhausen's *Factors of Evolution* has met with considerable approval in the Department of Biology of the USSR Academy of Sciences, and in the Scientific Council of the Moscow University, suggests that something is not right in some institutes of the USSR Academy of Sciences, as well as at the Moscow University.[4]

One cannot help voicing one's astonishment at the behavior at this session of Academician V. S. Nemchinov, who made approving remarks about the Mendelists-Morganists who spoke here, and who was indignant at their being criticized by representatives of the Michurinist trend, particularly Academician Lysenko. It is known that at the Timiryazev Agricultural Academy not everything is as it should be concerning the propaganda of Michurin's teaching. For many years, Mendelist-Morganist ideas have been propagated from the Chair of Genetics of this largest agricultural college in the country. I trust that Academician V. S. Nemchinov will appear at this session and will say clearly what trend in biological science will be followed in the future by the Timiryazev Academy. *(Noise in the audience. Exclamations:* Right! True! *Applause)*

The richest practice of the Socialist agricultural economy; the numerous army of vanguard workers—heroes of Socialist labor; and the accomplishments of our scientists all confirm with exceptional force that the Michurinist trend in biological science is the most advanced trend, arming practitioners for a fight to obtain high crop yields, to raise Socialist animal husbandry, to secure an abundance of products of agriculture.

The Michurinist trend in biology has opened ways of acclimatization of plants necessary for the national economy, and methods of advancing plants into new regions. The measures undertaken by our country to develop seed-cultures; to raise the fertility of the fields of Soviet and collective farms; to establish field-protecting forest plantings; to insure high, stable yields in the years of drought; these measures, as well as others, are unthinkable without a further, manifold development of the teaching of Williams-Michurin-Lysenko.

In light of the report read to us by T. D. Lysenko and its discussion at the session of the Academy, exceptionally urgent significance accrues to the problems of the right direction of scientific research and of the principles of instruction in biological sciences in higher education. These problems have a direct bearing on the shaping of a scientific world-view of our contingents. In scientific research and in the teaching of biology, the Michurinist trend

[4] This is evidence for the beginning of a heresy hunt.

must reign alone. The teaching of the gramineous system of agriculture, as applied to the peculiarities of different agricultural zones, must become the property of our agronomic contingents, and must be propagandized daily from the chairs of institutions of higher education.

We must put an end as quickly as possible to the situation in which the higher agricultural schools, including the Timiryazev Agricultural Academy, sometimes graduate agronomists who have learned well Mendel's "peas' laws," as Timiryazev used to call them, but who have only a vague idea of the creative methods of transforming the nature of plants and creating new strains, methods which have been worked out by Michurin and are being developed by Lysenko.

In conclusion, permit me to express my assurance that the All-Union Academy of Agricultural Sciences, which bears the name of the great Lenin, now that its staff is complemented by a new contingent of active members, will become a powerful center of development of the advanced agronomical science in our country, and that it will always carry on its banner the counsel of Comrade Stalin: "Science is called science precisely because it recognizes no fetishes, is not afraid to raise its hand against what is old and obsolescent, and listens alertly to the voice of practice and experience." *(Prolonged applause)*

4

Professor D. A. KISLOVSKI

K. A. Timiryazev Academy of Agriculture

In this discussion, we are essentially conducting a debate about the present situation of more than biological science alone. The essence of this argument lies in the question of methodological principles in treating scientific problems in any branch of science under Socialism. On what philosophical foundation grew up the bourgeois natural sciences? On metaphysics. The basis of experimentation lay in dismemberment, in concentration upon separate, particular problems, and in a loss of perspective that would have allowed a correct evaluation of the significance of these particularities. Hence the road led naturally to purely formalistic constructions and to an impotence in the face of solving actual problems.

Materialistic dialectics, which is the basis of our Socialist, scientific thought, demands a thorough cognition of objects, "a many-sided estimation of interrelationships in their concrete development" (Lenin, Vol. 26, p. 132). The Michurinist trend in genetics, which is being developed by T. D. Lysenko, is based on the dialectical method, and the reason for its theoretical value and practical efficiency is that it takes account of a biological object in its motion, its development, its conditioning by environmental phenomena.

Typical of the opponents of Academician T. D. Lysenko is their haughty attitude toward practice. They are trying to instill their inventions (like B. M. Zavadovski's artificial molting of geese) into practice with great insistence, without taking into consideration the actual need of practice itself, as it develops. Academician T. D. Lysenko had soundly estimated the needs of the growing Socialist economy, and managed to become a scientific guide of multimillioned masses in raising productivity of our agriculture. His ideas have become a material force, inasmuch as they took hold of the masses. This is why we must in every possible way support and develop the teachings of Michurin and Lysenko, must translate these ideas into practice of zoötechnical work.

I am glad to appear today in the same position concerning the basic problems of genetics and selection for which I had fought, together with other Michurinists, as far back as the discussion of 1936. I had reached this position quite a while back, on the basis of my own works, for which I had been labeled by the formalists as an "antigeneticist." By the way, let me say that it would be interesting to know with whom did Comrade Shaumyan confuse me when, in his published speech, he asserts that Kislovski is an advocate of a primary fund of genes that only become recombined?

Opponents of the Michurinist trend were trying to convince us that Academician T. D. Lysenko is intolerant of all criticism. Whereas I do want to criticize him and am profoundly convinced that he would be tolerant and attentive to businesslike criticism. He is intolerant only when it comes to a fight against metaphysics. Here there cannot be, and should not be, any compromise.

Academician T. D. Lysenko misuses our zoötechnical term "breed." Breed is a group of individuals whose heredity is far from being identical. Breed is a whole group of animals, whose qualities differ in many respects, and this is not accidental, but a distinctly regular phenomenon which is quite important. In spite of all the assertions of the formalists, a zoötechnician must deliberately create this qualitative variability, this hereditary differentiation within the limits of a given breed, while at the same time he must take into account and keep up the connection between parts of the breed and the breed as a whole. The essence of a breed is that which determines that it is a breed, and not an accidental aggregate of individuals. The essence of a breed is not in heredity per se (or even as conditioned by external circumstances only), but in the totality of the breed, created by human labor. The concept of a breed is not biological. In wild nature there are no breeds. A breed is a product and means of production.

It goes somewhat against my grain to hear that the basic task of the Academy of Agricultural Sciences is agrobiology. As I see it, its basic task is to develop from all angles the theory of agricultural production under Socialism. This implies a considerable obligation. Here agrobiology must play the role of only one means toward raising the level of production.

In conclusion, I would like to pause on two instances from the realm of development of Socialist, animal husbandry. The first is the need of a more specialized, individual approach to each animal. We could fully evaluate the enormous importance of this factor only after we saw the scope of the Stakhanovite movement. Here, the Michurinist genetics ought to be of help in many respects.

The second instance is raising not isolated animals, but organized, total groups—breeds. Unfortunately this side of organization of zoötechnical production did not, until the present, receive sufficient attention. *(Applause)*

5

Academician I. TH. VASILENKO

The founder of the science of agricultural machinery is Academician V. P. Goryachkin. His followers—workers in scientific-research and educational institutes—raised the science of agricultural machinery and tractors to a high theoretical level. We may rightfully feel proud of the fact that Soviet science has priority in the theory of agricultural machinery and tractors.

The peculiar feature of the science of agricultural machinery is the fact that it cannot be based merely on technical calculations, but must absolutely proceed from the requirements of agrobiology. Soil, plant, seed, as Academician Lysenko noted in his report, are living bodies. Therefore, only a profound interrelation between the science of agricultural machinery and agrobiology can create a true theory of these machines, and throw light on the path along which the construction of these machines has developed as well as on the methods of using them.

We consider that our science of agricultural machinery must be built on the foundation of agrobiology and technical sciences. It must be based, on the one hand, on the work of Timiryazev, Dokuchayev, Williams, Kostychev, Michurin, and Lysenko; on the other hand, on the work of Academician Goryachkin and his followers. This point of view has been embodied in the organization of new programs of study for the Molotov Moscow Institute of Mechanization and Electrification of Agriculture; and the All-Union Conference of representatives of the leading departments of the higher educational institutions dealing with the mechanization of agriculture has adopted this program for all faculties and higher institutions of learning devoted to this special field.

What are the demands that agrobiological science makes on the construction of agricultural machinery? How do these machines created in our land differ from machines meant to serve the same purpose in capitalist countries? Let us take machinery used for cultivating the soil. The task of plowing is to turn over layers of soil and to cut them into small particles. But Soviet agrobiology makes additional demands: not merely to turn over the layers, but to do it in such a way that the top layers of the soil, the texture of which has been broken up, should land at the bottom of the furrow, together with the weeds; that the texture of the soil should be improved and the weeds destroyed. To accomplish this the colter has been invented. The cutting equipment of our plows is projected on the basis of the theory founded by Academician Goryachkin. At the present time we construct only plows equipped with colters, and our plows stand on a high level with respect to their agrotechnical qualities.

Conditions are quite different with regard to plowing implements in capitalist countries. More than two thousand different types of plows are in use in all parts of the world. This variety of types of plows has no scientific basis and is the result of competition between various firms. To eliminate this confusion in the construction of plows, a special government commission was formed in the U. S. A. which included in its membership agronomists, engineers, even mathematicians. The mathematicians suggested the surface shape of the cutting equipment of the plows. The commission achieved certain results with regard to the technique of construction. However, the agronomy of the U. S. A. proved powerless in the matter of gaining ground for those requirements which determine a high agrotechnical level in plowing. Colters are not used on plows in the U. S. A. and that leads to terrific weed-clogging of the soil. In the U. S. A. agronomy is not tied up with biology. No agrobiological science exists there. As for biology, in the U. S. A. it is completely detached from the interests of the people and is reactionary in its essence.[5]

Agronomy and biology in the U. S. A. have permitted a predatory attitude toward the fertility of the soil in the very process of plowing. These predatory tendencies found even stronger expression in the construction of the grain-combine. There they apply the system of cutting off the plants at a point high from the ground and of gathering in only the grain. The straw, however, is scattered all over the field by means of specially adjusted rotating devices. With the chaff and the straw, a large quantity of seeds of weeds is spread over the fields. Softening up and clearing of the stubble field after the harvest is almost never practiced.

The U. S. A. occupies one of the first places in the world with regard to the problem of weed-clogged soil. There the yield from the fields is not on a high level and is not increasing.[6] Even in the American press, voices are raised to the effect that the soils are exhausted to the limit, that the farmers are unable to obtain good results from worn-out earth and are being ruined. Biology in that country watches indifferently how greedy purveyors of agricultural technique force the farmer to despoil fertility.

Soviet agrobiological science makes high demands on the quality of work rendered by a combine. The combine must guarantee a high agrotechnical level in the matter of harvesting, i.e., cutting of crops close to the ground, collecting of straw, chaff, and the seeds of weeds. One cannot make successful use of plows with colters if the combines leave high stubble; it clogs the colter, and the operator of the tractor is obliged to stop the tractor frequently. The result is a waste of fuel which sometimes reaches 20 to 25%.

To assure cutting of crops close to the ground with complete grasping by the header, the combine must have a sufficiently effective thresher and a powerful engine. We have solved this problem by the creation of the combine "Stalinets-6," which can with the same grasp handle 25% more growing matter than the former combine. In the motorized combine the thresher has been widened, the straw shaker has been lengthened, and a separator with powerful action has been constructed. To stop the scattering

[5] This is called to the attention of the United States Department of Agriculture.
[6] This accounts for our crop surplusage.

of the seeds of weeds, a straw and chaff stacker has been built onto the trailer combine.

Since we have combines which assure close cutting of crops and have stackers connected with them, also disk plow-harrows and plows with colters, we can mechanize in full all operations of harvesting and autumn cultivation of the soil, as demanded by Academician V. R. Williams. In 1947, elaborate tests were conducted on an acreage of 10 thousand hectares in which harvester combines and machinery for autumn cultivation of the soil were used. The tests gave completely satisfactory results; the collective farmers (kolkhozniki) were very well satisfied with this kind of work.

At the same time that we are creating machines for the harvesting of crops we are preparing a solution of another most important problem: the mechanization of the crops of perennial plants. The stacker, the pick-up machine, the drag harrow for the gathering of straw are also suitable for the harvesting of hay. Our system of machine-construction has produced good hay mowers of the horse-drawn, tractor, trailer, and attached types. In addition, a motorized hay-mowing machine of original Soviet construction has been produced, and it is very easy to maneuver. It has a speed of 18 kilometers per hour.

In creating all these machines, the Soviet science of agricultural machinery opens wide possibilities for the application of a progressive system of agriculture on our fields.

I shall now speak of machines for the mechanization of cattle-breeding farms. Capitalist technique has produced milking machines. They operate on a two-stroke principle. It has been discovered that these machines are harmless only for cows with healthy udders. In the case of cows suffering from hidden mastitis, machine-milking causes aggravation of the disease.[7] Our Soviet investigators have discovered the fact that, during milking, a cow's physiology requires intermissions between compression and suction of the teat. During these intermissions the teat rests and its blood circulation is restored, similar to the process when the calf sucks the cow. On this principle we have created a milking machine which operates on a three-stroke basis: we have alternating strokes of compression, suction, and resting of the teat. In comparison tests this milking machine has taken first place, and this is the result of consideration for the demands of agrobiology in the creation of machines for purposes of this kind.

However, we do not always succeed in carrying out the demands of the science of agrobiology. We are behind in the mechanization of the processes of cleaning and sorting grain and

[7] Are cows with diseased udders milked in Russia? Who drinks the milk?

seeds. It has been proved that when seeds with higher specific gravity are planted the harvest is sometimes increased up to 25%. To date, we have no machines for the sorting of grain and seeds according to a given characteristic. In this respect there is before us constructors a clear and concrete task, which in itself is a guarantee that such a machine will be built.

The examples which we have quoted prove convincingly that the path which we have chosen for the development of our science of agricultural machinery construction is the right one. We already have a number of constructions in the line of agricultural machinery and tractors which have been awarded the Stalin Prize.

Based on Michurin agrobiology, we will continue to raise the theoretical level of our science, to guard its priority, and to intensify its practical trend toward solving the most essential problems of increasing crops, of raising the productivity of labor, of creating an abundance of foodstuffs, and of strengthening the welfare of our people. *(Applause)*

6

Academician A. N. KOSTYAKOV

Academician T. D. Lysenko in his report has brilliantly developed the basic scientific ideas of Michurin and Williams, and has demonstrated their enormous importance for the entire biological science. I shall pause on the significance of these ideas in the realm of agricultural reclamation, directed toward improving the unfavorable natural conditions.

V. R. Williams, in the course of his entire scientific activity, came in direct contact with the problems of reclamation. And he, by right, may be considered the founder of our science of reclamation. Works by V. R. Williams offer a great deal for the development of the entire science of reclamation and of meliorative production. V. R. Williams has developed and established the law of irreplaceability, of equivalence of factors of plant life, and has shown its importance in agricultural economy. This law touches upon the fundamental aspect of reclamations; it says that acting on water factor alone is insufficient in reclamation; that simultaneously with changes in water conditions, and corresponding in quality and quantity to these changes, attention must be paid also to changes in other factors of plant life—food, heat, etc. V. R. Williams used to say: We note two elements of soil fertility—water and elements of plant food. Both elements of fertility—water and food—are absolutely necessary for the plant; neither of them can be replaced by another.[8]

Hence the conclusion: all water utilization in our reclamation systems, all their exploitation every year over any area, must be tied up with agrotechnics, tied up with the degree of soil fertility and the condition of plants. In every reclamation project there must be worked out not only the meliorative, hydrotechnical measures, but also the agricultural measures: crop rotations, fertilizers, soil cultivation, etc., indissolubly tied up into a single complex. Reclamations must be calculated exactly for a definite type of economic utilization of reclaimed areas, following planned tasks of the state. The law of equivalence of factors in agricultural economy poses fully determined tasks both in exploitation, and in planning of reclamations.

Failure to follow this law leads to a break between the construction and assimilation of systems, to lowering of crop yields, and lesser effectiveness of reclamations. By changing the water regime of the soil, reclamations substantially affect the character of processes going on in the soil, and change these processes. Therefore, it is necessary so to direct reclamation measures that all the processes, all the changes occurring in the soil, have a positive character, necessary for our economy. It is necessary by means of reclamation to control the dynamics, not of the water regime of the soil alone, but also of its atmospheric and nutritional regime, in the interests of increasing fertility and obtaining high crop yields. The teaching of V. R. Williams gives us the key for the correct solution of these problems.

Highly important theses and problems for the business of reclamation follow from works of V. R. Williams, dedicated to the teaching of creating soil structures and of the grass-field system of agriculture. This teaching has a fundamental significance in all problems of reclamation. But it assumes particular importance in problems of irrigation, since they have to be solved under the conditions of dry steppes, the soils of which are marked by a lack of structure.

On irrigated lands, the grass-field system of agriculture permits insuring the structure of irrigated soils by transforming these soils. When the irrigated soil has lumpy structure, the watering and irrigational norms are lowered considerably and the watering regime changes—it becomes less intensive. Moreover, only in the

[8] It will be interesting to discover if this bit of common knowledge will be claimed as a Russian "first."

presence of lumpy structure is it possible to regulate the necessary concentration of soil solutions, and to protect soils from both insalination and marshiness. A soil with a lumpy structure removes the antagonism between water and nutrient substances and creates conditions for a full manifestation of effective fertility. Only in a lumpy soil does any plant-breeding measure—irrigation, fertilization, seed selection, vernalization, etc.—manifest its full effectiveness.

At the same time, introduction of a grass-field system of agriculture presents its own demands to the regime and technique of irrigation. V. R. Williams considered that it was necessary to put in effect a radical reorganization of the technique of irrigation. He was an adherent of artificial rainfall. But artificial rainfall alone he regarded as insufficient, and he posed the problem of a fundamental reconstruction of the entire irrigational network on new, rational principles. Besides maintenance of structure of irrigated soils, V. R. Williams, developing further the ideas of V. V. Dokuchayev, considered it necessary to form on the irrigated areas forest zones for wind protection, attributing to them extremely great importance. Forest zones permit a favorable change in the climatic conditions of the irrigated territory.

In connection with working out the problem of grass-field system of agriculture and forming forest zones, V. R. Williams presented an idea of the necessity of systematic distribution over the country's territory, along the basic relief elements (watershed, slope, etc.), of the basic cultivation types (forest, field, meadow), and of a tie between the agricultural and water economies of the country. Developing V. V. Dokuchayev's ideas, V. R. Williams speaks of a broad, complex tie-up between hydrological, reclamational, forestry, and agricultural measures.

All these measures, directed toward a fight against drought, toward soil conservation against erosion, toward a rational utilization of water resources, must guarantee changes not only in the nature of the soils, but also changes in the hydrological, and even climatic, conditions of separate regions. Realization of these grandiose tasks is possible only in our country with its Socialist agricultural economy, in the presence of planned organization of the entire national economy.

Such are the basic theses and tasks following from V. R. Williams' teaching. It is necessary to utilize this teaching even deeper, even fuller and wider, to develop it further, and to incorporate it in our reclamation practice. The teaching of I. V. Michurin and V. R. Williams gives us methods of changing the nature of plants, animals, nature of soils, ways to change hydrological and climatic conditions. Using this teaching, our Soviet science will achieve new, great successes on the path of nature transformation and will bring a great contribution to the task of building Communism in our Country. *(Applause)*

7

Academician P. P. LOBANOV

Comrades, the problem concerning the state of biology discussed at this session is of exceptional importance, not only to the further development of biology and agriculture, but also to the further growth of the Socialist agricultural economy with which this session is indissolubly connected.

Our agriculture, developed upon the doctrine of the most victorious teachings of Marx, Engels, Lenin, and Stalin, must aid daily in the struggle for the further progress of our agricultural production. *(Applause)* It must also help in a successful solution of those tasks which are put to Socialist agriculture by our Party, by the Soviet Government, and by Comrade Stalin personally.

Such aid can be rendered only by a progressive agrobiological science standing at par with the olympian tasks of a socialistic system of an economy possessing inexhaustible powers and opportunities present in a collective-farm (kolkhoz) regime.

Such a progressive agrobiological science is the Soviet Michurin agrobiology, the foundations of which were laid by Michurin and Williams and which is creatively developed by Academician T. D. Lysenko.

The Michurin doctrine of recasting the nature of plants is organically allied with Williams' theory concerning the soil-forming process, the active effect of man on the soil, the restoration of soil fertility and increased yield.

Michurin's theory opens a new and superior stage in the development of materialistic biology. It makes it possible, not only to explain the phenomena of life, but also to alter the life of plants and animals in a direction necessary and useful to man.

The Michurin doctrine is a socialization and a powerful development of all the best that has been accumulated in the past by science and centuries-old practical experience in the understanding of the nature of plants and soil and in the perception of agriculture.

Our scientists, appearing as the real representatives of this science, and who contributed greatly to the elimination of the centuries-old backwardness of agriculture, aid in the organization and the strengthening of the economy of the collective farms, the MTS (Machine and Tractor Service Station),[9] the state farms; and also render a great service in the postwar restoration and further development of agriculture.

But the successes thus attained in the progress of agriculture and in the development of agricultural science do not satisfy us. These successes can be regarded only as a substantial premise for a further vigorous improvement in agriculture, cattle-breeding, and the other branches of husbandry, and as a basis for the future growth of scientific thought. We have at our command enormous, inexhaustible reserves for an increase in the output of agricultural products, also sufficient resources for a rise in the productive power of agricultural labor. It suffices to familiarize oneself with the results of the work of the progressive collective farms, the MTS, the state farms, and the innovators of agriculture in order to see how large these reserves are. The successes attained by the progressive economy are accessible to everybody.

Adoption of the technique in our economy can and must lead to a manifold rise in the productivity of our agriculture. Progressive Soviet agrobiology led to the building up of a new breed of Kostroma cattle, rich in milk and considered one of the best in the world. We cannot agree with the spokesmen of so-called formal genetics who assert that the Kostroma cows had long been endowed with genes which were revealed only lately. Yes, if one should be guided by it and remain inactive, then, of course, organization in quality and breed assortment will become extinct or will be retarded as is predicted by Academician Schmalhausen.

Owing to the daily assistance and solicitude by the Soviet Government concerning the strengthening and the development of the collective-farm system, our agriculture attained large successes in the postwar period.

In agriculture there are grave premises for a further vigorous rise in production, an increase in cultivation, and a growth in animal husbandry. This fact again and again bears witness to the advantages of a socialistic economic system, to the inexhaustible strength and possibilities of the collective-farm regime. The chances for raising production are guaranteed by an ever growing mustering of energy in agriculture, the growth and further improvement of the MTS, by the increasing number of highly qualified personnel, and by the unified aspiration of the Soviet people

[9] The MTS' services to the collective farms consist in plowing, tractor work, sowing, harrowing, threshing, and so forth.

to develop steadfastly the most progressive Socialist agriculture in the world.

The most important proviso of our successes in the postwar rise of agriculture is the development of our progressive agronomic science and the inculcation of its achievements into agricultural production.

The teachings of Michurin and Williams are growing continuously, and credit for this development should be given to Academician Lysenko. He skillfully tied the Timiryazev-Michurin theory of form development and nature changes in plants and animals to the Dokuchayev-Kostychev-Williams theory of soil formation and methods of increasing soil fertility, arming the collective and state farms in the battle for continuous increase in yield and the improvements of our agriculture.

Academician T. D. Lysenko and his followers promote, not in words, but in fact, the best traditions of our agrobiological science, and have already achieved large practical improvements.

The ranks of Michurin followers, progressive Soviet biologists and agronomists, grow constantly. The outstanding successes of our Soviet plant culturists and selectionists—such as Yakovlev, Kanash, Lukyanenko, Zhdanov, Ushakova, and also of such cattle breeders as Yudin, Greben, Shteiman, Balmont, Philyanski, Vasiliev, and many others—are a guarantee that breed and quality organization in our country will not only survive, but will grow more and more.

These achievements of the true followers of Michurin and Lysenko were graphically demonstrated beforehand in a series of speeches in the report. These articles elucidated the important results of our progressive scientific research institutions and of the individual scientific and production workers. The results of these labors prove not only the scientific correctness and practical efficacy of the basic theses of the Michurin genetics as a progressive trend of Soviet biology, but also contribute greatly to the further growth of this trend.

The opponents of the Michurin school, contrary to his adherents, came to the session equipped merely with promises of future "great" discoveries, something we have heard many times in the past. Speaking frankly, the opponents of the Michurin theory arrived without any real, practically tangible results.

And this is one of the incontrovertible evidences of the defect in their theory.

The main tasks of agriculture at present are to increase all kinds of harvests and to increase the total yield of agricultural products; to heighten the cultivation of the soil by inculcating the achievements of progressive agronomic science; to develop all kinds of animal husbandry, increasing its productivity; to strengthen further the economic organization of collective farms; and to provide an all-embracing development of their social economy.

In view of these tasks we cannot regard as satisfactory the rate of increase of yield already achieved, although it has been considerable. In the light of the tasks still to follow, we cannot be satisfied with the tempo and the method of scientific work which up to the present are being used by many of our scientists in the treatment of agricultural problems. Now has come the time when we must reach a radical increase of yield in our country. It is the duty of the scientists to adopt such measures as would prove an aid to all the collective and state farms in increasing the yield to a degree exceeding all previous records and which would assure a further sharp increase in the productivity of labor. This is indispensable in satisfying the renewed demands which at the present time must be met by agriculture in our Fatherland.

Considering the new tasks which agriculture must solve, we cannot restrict ourselves to the accomplishments of individual agrotechnical methods. It is necessary to inculcate on a large scale an entire complex of scientific agrotechnical measures. A vigorous means for a radical strengthening of agriculture and an increase in the development of agriculture is a daring implantation and realization in practice of a herbicultural system of agriculture developed by the coryphaei of Russian agronomic science, Kostychev, Dokuchayev, and Williams. That which has been an eternal dream of progressive Russian agronomists must now be transformed into a concrete program of action and must become in the near future a reality.

Socialist agriculture has at its disposal at the present time all that is necessary to transform into life this advanced method of agriculture.

It is our duty to aid the collective farms, the MTS, and the state farms in the realization of an herbicultural system of agriculture. While doing this, we must provide for further concretization of this plan, making it suitable to individual areas of our country.

From the Soviet scientists our country expects work which will result in rich and stable harvests in all the zones of the Soviet Union, as well as in high productivity in the field of animal husbandry and all the other branches of our economy.

The workers on the collective farms, men and women alike, the Heroes of Soviet Labor, the Laureates of the Stalin Prizes, the remarkable masters of high yields, the organizers of collective-farm production, having successfully utilized the achievements of advanced science, became the outstanding people in agriculture and the true leaders who turned advanced science into practice. The experiments of the progressives and their accomplishments enrich our agricultural science, supply it with new strength, and open before it new vistas.

Our science is wholly at the service of the Soviet Government and the Soviet people.

The Soviet people, from whose ranks our scientists rose, demand from them supreme fidelity to the great ideas and to the party of Lenin-Stalin, and expect from them the development of science in a direction responding to the vital interests of a Socialist society and the unification of scientific forces of the various branches of science for the rapid achievement of the aim in view, namely, the erection of a Communist society.

Our science cannot tolerate lack of ideas, scholasticism, and metaphysics.

In agriculture, which deals with plants and animal organisms, biology is of utmost importance as a science which makes a study of conformity in the development of live organisms.

Biology can prove an invaluable aid in practice on condition, however, that it springs from materialistic dialectics and creatively develops the theoretical heritage of such coryphaei of our native biological science as were Timiryazev and Michurin.

However, in the field of biology, not all the scientists follow the road taken by the foremost Russian biologists. Side by side with the advanced, materialistic, Michurin school there is another, diametrically opposed, anti-Michurin theory, in reality a reactionary, idealistic drift in science.

The Michurin trend of biology is founded on materialistic ground, and maintains its belief that a change in conditions of life is inevitably followed by a change in the nature and the form of organisms. The nature of organisms is shaped by living conditions and changes with living conditions.

The Michurin school of thought is based on the recognition of the fact that new attributes of organisms acquired by them in the course of their development under the influence of changed conditions in external environment are transmitted to and inherited by their offspring.

The Michurin theory supplies the practical worker with scientific methods of transmutation and improvement of the nature of plants and animals in a direction beneficial to man.

The most consistent partisan of the advanced, progressive, Michurin trend in biology is found in the person of Academician Lysenko. His great service is in the fact that he carries on a brave struggle based on his principles against the idealistic dogmas of the Mendel-Morgan genetics (Weismannism), against the metaphysical and formalistic theses in biology, and thus contributes to the rapid growth of the Michurin doctrine as a new stage in materialistic biology.

And it is not merely by chance that the champions of the Mendel-Morgan theory conceal Michurin's studies, lower the enormous theoretical and practical importance of his efforts, and at the same time in every way push forward the so-called scientific advancements of Mendelism-Morganism. For instance, such a stand has been taken by Academician Schmalhausen up to the present time. How else can it be explained that in his work *Factors of Evolution,* published in 1946, he never mentions the names of such giants of scientific thought as Michurin and Timiryazev, who dedicated their entire creative lives to these problems.

Academician Schmalhausen replaces the conformities of living nature with formalistic schemes leading to antiscientific deductions of would-be, inevitably arising, standstills in evolution.

This is the logical end reached by researchers who have lost contact with life and with advanced science.

It is clear to everyone that the views of Academician Schmalhausen and his adherents impair our agriculture as they poison the conscience of the practical workers in agriculture and that of the students in our schools with idealistic metaphysics, disarm the agronomists, the zoötechnicians, and the other specialists in agriculture.

Such a pseudoscientific stand begets enmity and directly ignores the remarkable achievements of the Michurin adherents and, above all, of Academician Lysenko.

Formal geneticists have dedicated themselves, as was shown by their speeches at the session, to the defense of reactionary theories.

They exert themselves, up to the present day, to the end of tearing Lysenko away from Michurin. Similar attempts to set off the Michurin theory against the views and convictions of Lysenko, who creatively developed this theory, are using worthless means and should be condemned by us. *(Applause)*

Attempts to reconcile two different trends in biology, to unify the reactionary trend in biology with the advanced and progressive Michurin school or to adopt some middle-road policy as proposed by Academician Zavadovski, are inevitably doomed to failure.

Creatively developing and advancing biological theory, Academician Lysenko is linked daily and closely with the collective and state farms. We, the workers in the Department, recognize more than anyone else what an enormous service is rendered to the practical worker by Academician Lysenko *(applause)* in organizing their forces toward a rapid implantation in production of the conquests of advanced science. His activities are a good example to each Soviet scientist, as only close accord with practice can result in successful growth of progressive science. *(Applause)*

The anti-Michurin trend in biology is founded on metaphysics and negates the possibility of quantitative changes in the inherited properties of plants and animals under the influence of changes in living conditions. It denies the possibility that living organisms can inherit attributes and qualities arising in the living body under the influence of living conditions. The anti-Michurin trend springs from the false assumption that man is unable to perceive the causes and direct changes in heredity.

A thoroughly reactionary and mystic view disarms the practical worker in his labor on the cultivation of new kinds of plants and new breeds of animals. It reduces the role of the selectionist to the role of the treasure-hunter, to the role of a man passively awaiting the appearance of desired forms of plants and animals. The champions of this school are prisoners of the bourgeois theories of Morgan, Mendel, and Weismann. They have not yet freed themselves of their servility to bourgeois science and, willingly or unwillingly, choke science with antiscientific ideas.

The adherents of the anti-Michurin trend in biology, referring frequently to Darwin, forget that the Darwin theory explains only the evolution of the organic world, while Michurin and his followers constantly teach how to change as well as create systematically new forms, how to alter the nature of plants and animals in a direction useful to man.

The biologists who follow the anti-Michurin theory have lost contact with life and are impractical. They cannot with their fruitless researches be of any help to the collective and state farms.

We are not aware of any valuable results obtained by the formal geneticists on the basis of their work. And a science which does not aid production, which does not help the practical worker, and which does not assist the Soviet people in their aim to build a better life, is not our science! *(Applause)*

And so the anti-Michurin trend in biology was correctly and deservedly subjected to severe criticism by the scientists present at the debates of this session.

The speeches by the representatives of formal genetics showed that they still adhere to previous reactionary stands and evidently intend to be a drag in the development of advanced biology.

Such an unenviable role was played by the group of formal

geneticists who renounced the scientific principles of materialistic dialectics.

The representatives of formal genetics did not perceive the worthlessness of their theory, lacked the courage to see the error of their viewpoint, did not arrive at any deductions from the speeches at this session by many scientists and practical workers—the representatives of the Department—who reflected not only their personal viewpoint but also that of the largest collective body of scientific workers and specialists in agriculture and collective farming. *(Applause)*

In his speech yesterday, Academician Zhukovski clearly defined his attitude toward Mendelism. He even appealed to the Soviet biologists to bow before Mendel, thus swearing to the correctness of his teachings.

To your appeal, Academician Zhukovski, for us to rally to the Mendel theory, we reply as follows: If you will continue to follow the old road of Mendelism, we must part our ways. *(Applause)*

We, the Michurin adherents, rejecting together with the collective farms, the reactionary, idealistic biology, will under the banner of an advanced, progressive biology and with our Lysenko fight for the further blossoming of materialistic biology and for more and more scientific aid to the state and collective farms. *(Applause)*

In view of the enormous tasks to be met by our Academy, problems of scientific propaganda and of further development of the Michurin theory acquire a special meaning.

It is necessary to expand by all means possible scientific research work directed toward the development of essential problems of the Michurin school in close connection with a solution of the important questions in agriculture.

We must change radically the methods of teaching selection and genetics in the higher and lower agricultural educational institutions, basing the methods of training of these branches on the Michurin theory.

A great role in the correct education and formation of staffs of biologists, selectionists, agrotechnicians, zoötechnicians, and other specialists in agriculture will be played by education on a level with contemporary science and the great precepts of Michurin, Williams, Lysenko, and other progressive scientists.

The Soviet scientists completely devoted to their people, equipped with the most progressive doctrine in the world—the Marx-Engels-Lenin-Stalin doctrine—will secure the further growth and transformation into life of the advanced Michurin-Williams-Lysenko school.

Undoubtedly our scientists facing, as they do, the constant

cares of our Party and Government will in the near future reach new important successes which will contribute further to the vigorous flowering of our socialistic agriculture; thus they justify the high trust in them of the Soviet people and its great leader, Iosif Vissarionovich Stalin. *(Wild applause)*

8

Academician V. S. NEMCHINOV

The principal workers of the Timiryazev Agricultural Academy, leaders of its most important departments, Academicians I. V. Yakushkin, V. P. Bushinski, P. M. Zhukovski, A. R. Zhebrak, and Professor D. A. Kislovski, have already appeared at this session.

They differ in their views on many questions of the current scientific discussion. Inasmuch as all the professors of the Timiryazev Academy are unanimous concerning the basic question of the tasks of agronomical science—namely, to transform nature in the interests of Socialist construction in our country—I, as the director of the Timiryazev Academy, see nothing reprehensible in these divergences of views on some scientific questions.

No one has contested here that I. V. Yakushkin and V. P. Bushinski are fully for the positions of the Michurinist trend in biology. I believe that the scientific position of Professor N. N. Timofeev, who is in charge of the Chair of Selection of Fruit and Vegetable Cultures at the Timiryazev Academy, is also recognized by all those present here as being in full accord with the advanced, Michurinist trend in biology. Since the death of Academician P. I. Lisitzyn, the Chair of Selection of Grain Cultures at the Timiryazev Academy has been the charge of Academician P. N. Konstantinov, a man who gave our collective-farm fields many excellent strains of grain cultures now occupying dozens of millions of hectares. He showed by his deeds that his scientific views help him to work creatively for the good of Socialist construction.

The course on Darwinism at the Timiryazev Academy is conducted by Professor A. A. Paramonov, author of a textbook on Darwinism, an excellent lecturer, a scientist standing completely on the positions of Darwin and Timiryazev. He has differences with T. D. Lysenko concerning particular scientific questions, but it is a lie that A. A. Paramonov in his lectures "reviles" T. D.

Lysenko. At the Timiryazev Academy, in charge of principal chairs, are working such recognized disciples of V. R. Williams as V. P. Bushinski and M. G. Chizhevski. And Williams' teaching is inseparably intertwined with the Michurinist trend in biology.

Thus, there remains only the accusation connected with Professor A. R. Zhebrak.

The scientific public opinion at the Timiryazev Academy has sharply condemned the antipatriotic action of A. R. Zhebrak. I, as the director of the Academy, was among the first to respond, on the pages of the *Literary Gazette,* to the article of our writers about A. R. Zhebrak.

Nevertheless, the Timiryazev Academy, having sharply condemned A. R. Zhebrak's behavior, did not forget that he had come to the Academy from the ranks of farm hands, and had grown, within its walls, from an undergraduate and a postgraduate student to a professor of the Timiryazev Academy and an active member of the Academy of Sciences of White Russia.

I consider wrong the demands of some members of the current session of the Lenin Academy of Agricultural Sciences to put an end to A. R. Zhebrak's works in the domain of amphidiploidy and experimental polyploidy. These works of A. R. Zhebrak promise some highly useful results for our Socialist agricultural economy.

As for my attitude to T. D. Lysenko's reports, I agree with him in many respects. I agree with the basic direction of the report, which is to mobilize agronomical science for an active transformation of nature in the interests of the Socialist agricultural economy. I have vast respect for his ability to become the center of agronomical science that actively involves the collective-farm masses in creative scientific and practical work on the fields of the collective farms. However, I do not agree with his appraisal of the chromosomal theory of heredity which is the scientific golden fund of biology. We can transform nature, utilizing essentially the changing generations of plant and animal organic forms, by means of deliberate action upon the chromosomal complex; among means of action are also included methods of rearing, such as for instance, hardening.

I consider that the chromosomal theory of heredity has become a part of the golden fund of science of mankind, and I continue to maintain this point of view.

A voice from the audience: You're not a biologist; how can you pass judgment on it?

Academician Nemchinov: I'm not a biologist, but I have the opportunity to verify this biological point of view and theory from the standpoint of the science in which I conduct my scientific research, and in particular from the standpoint of statistics, as science. *(Noise in the audience. Applause)* From my point of view this theory also agrees with my concepts. But that is not the question. *(Noise in the audience.)*

A voice from the audience: What do you mean, it's not the question?

Academician Nemchinov: Very well, let that be the question. Then, in fact, I must say: I cannot share the point of view of those comrades who declare that chromosomes bear no relation whatever to the mechanisms of heredity. *(Noise in the audience.)*

A voice from the audience: There are no mechanisms.

Academician Nemchinov: That's how it seems to you, that there are no mechanisms. This mechanism cannot only be seen, but it can also be stained and determined. *(Noise in the audience)*

A voice from the audience: Yes, there are stains. And statistics.

Academician Nemchinov: I do not share the point of view expressed here by our esteemed chairman, Academician Lobanov, that the chromosomal theory of heredity, and in particular some of Mendel's laws, represent some sort of an idealistic point of view, some sort of a reactionary theory. Why, personally, for instance, I believe the reverse is true: I think such a thesis incorrect and that is my point of view, though it may be of no interest to most. *(Noise in the audience. Laughter.)*

A voice from the audience: It's of great interest.

Academician Nemchinov: I have never concealed this point of view. I think that the courses of genetics in our institutions of higher learning must present and develop the correct theses of T. D. Lysenko, but the principles of the chromosomal theory of heredity must not be kept hidden from the students.

Those who said here that I, as the director of the Timiryazev Academy, am responsible for its policy were right; I consider this policy to be the right one. *(Applause)*

9

V. N. STOLETOV

Deputy Director, Institute of Genetics of the Academy of Sciences, USSR

In the beginning, it seemed as if Professors of the K. A. Timiryazev Agricultural Academy in Moscow, A. R. Zhebrak and P. M. Zhukovski, had no desire to appear at this session.

Apparently, they wanted, as the saying is, to pass it in silence. But when this failed, they came out on the rostrum and began to read something like "educational" lectures. They pretended not to understand what the session is really about. Speech by Academician V. S. Nemchinov has made a great deal clear to us. It has become obvious to everyone that Academician Nemchinov, as director of the Academy, fully shares the point of view of his Professors-Morganists, and also apparently approves their behavior.

In the course of the debate, it had been noted that the leadership of the Timiryazev Academy shows no particular yearning to harbor within its walls scientific workers who are guided in their work by the Michurin doctrine.

Academician V. S. Nemchinov either denied the accusations lodged against him in this regard, or else placed the related facts in the category of errors. Naturally, a leader can commit errors. But there are errors and errors. There can be directed, deliberate errors. It seems to us that Michurinists feel ill at ease at the Timiryazev Academy precisely because of such intentional "errors" on the part of the leadership. Academician Nemchinov had expressed here his personal opinion concerning the chromosomal theory of heredity. One's personal opinion naturally becomes reflected, one way or another, in one's work. And this effect is what we observe here.

Timiryazevists, past and present, love "Timiryazevka." ["Timiryazevka" is a fond, colloquial abbreviation of the "Timiryazev Academy."] But they love the "Timiryazevka" of V. R. Williams, the "Timiryazevka" of M. F. Ivanov, and they have no respect for the "Timiryazevka" of A. R. Zhebrak, who is spreading the ideas of Morganism. *(Applause)* Evidently the majority of the Timiryazevists had heard with regret the statement of Academician Nemchinov that in his opinion the chromosomal theory has become a part of the golden fund of science.

Academician Nemchinov: Unquestionably, it has. Such is my opinion.

V. N. Stoletov: Whereas in our opinion, this "contribution" to science brings nothing but harm to the true, materialistic biology.

A voice from the audience: Right! Right!

V. N. Stoletov: Already in his first attacks (in 1935-36) against Mendelism-Morganism, Academician Lysenko had warned his listeners and readers that this was not the matter of some particular problems of biology, but of principles, of the direction of research in biological science. The basis of dissent among the biologists lies in their different views on the process of evolution in the plant and animal kingdoms. The character of a biologist's research and the success of this research depend a great deal on whether he occupies the scientific, materialistic position, or whether he is a prisoner of idealism.

Knowledge of the true laws of development of plant and animal forms is necessary to us for practical control over form-production, for creation of desired forms of plants and animals. Whereas success in the knowledge of the living apparently depends primarily on the scientific quality of thought. For Academician Lysenko, the fundamental aim of discussion was a defense of the scientific methods of thinking in the course of investigation of biological laws.

Moreover, he continued to regard as an indisputable axiom the thesis that the scientific quality of one's thought in biology must be defended not by scholastic arguments, but by a persistent study of objective reality, by a constant verification of the acquired knowledge about the laws of nature—knowledge to be tested in the fire of practice, of experience. Causality and necessity must be sought in objective reality; and the underlying principles, when found and discovered, must be immediately tested in the same objective reality, in an attempt at a better and better practical control over the living nature, in the interests of man.

The guiding ideas for working out a problem of such importance to the biological science as the causes of variability of organisms were sought and found by Academician Lysenko among the classics of Marxism. In one of his reports, he says that in the works of our masters of Marxism, one can gather not only the general, guiding ideas for the study of changing heredity, but even some direct indications as to where the changes come from, in what way they arise in the organisms.

Lysenko did not answer his opponents with a reference to general, guiding ideas. He answered them with the success of his practical research work; he answered them with his experiments, comprehensible, lucrative as to the conditions of their execution, profound as to their theoretical formulation. For Lysenko, the center of discussion was not in conference halls, but in the massifs of the collective and Soviet farms, in the greenhouses, in the experimental fields. He proved the scientific quality of his thought in the realm of biology by the increased crops of self-pollinators, after their intrastrain crossing; by the increased crops of cross-pollinated plants, following their free repollination; by the higher yield of potatoes in the south and the improvement of its stock qualities under the influence of summer plantings; by obtaining high millet crops; by dozens of facts of creating vegetative hybrids; by a directed transformation of any winter form into a hereditary spring form, and vice versa.

For Academician Lysenko, the years of discussions were filled with big research work on the key questions of genetics. But simultaneously during the same years, guided by the theories he was working out, he continued to solve large economic problems. In his case, solution of economic problems did not proceed parallel to theoretical works: it was intertwined with the latter, was based on them, issued from them.

This was not the way the Morganists behaved. It did not occur to them to study Lysenko's theses experimentally. On the contrary, they had mobilized all the genetic hypotheses and picked out all the falsely interpreted old facts; with all this insufficiently up-to-date arsenal, they came down upon Lysenko's positions. To the newer and newer verified facts advanced by Lysenko, the Morganists-Mendelists invariably opposed always the same old hypotheses and some few facts.

Apparently, our home-grown Mendelists-Morganists cared very little that the country expected from them some actual contribution to the biological science. Mendelists-Morganists began to attack Academician Lysenko solely for the sake of defending Weismannism.

In the course of discussion, Academician Lysenko and his followers have smashed the theoretical bases of Weismannism in our biology. All thinking people became convinced that Weismannism is foreign to our world view; and as for the practical aspect, our Weismannists, like the king in a well-known fairy tale, are naked.

The paths of the Michurinists and the Morganists have diverged into diametrically opposite directions.

If the present situation in biology were such that the divergences among the biologists concerned merely this or that particular scientific problem, then the question could be solved quite simply. It would have been possible to gather together Professor A. R. Zhebrak, Academician I. I. Schmalhausen, Professor P. P. Dubinin, and other Morganists-Mendelists, and take them for a drive all over our research institutes, where the Michurinists are at work, and over our selectionist stations, and to demonstrate to them in numerous experiments the scientific truth and force of the Michurin doctrine. They could have been shown the real vegetative hybrids which help to understand correctly, in the Michurinist fashion, the essence of heredity and its mutability. At present it is already possible to demonstrate that there is not a single grafting—of course, if it is done the right way—that does not produce changes. It would be possible to show many dozens of spring forms obtained from winter ones, and winter forms, obtained from the spring ones.

On the way from one institute to another, it would have been possible to drop in to visit the strain-testing sectors and take a look here at the results of testing the strains obtained via directed changes. In their quality, these new strains beat the old strains. Furthermore, experiments conducted by the Michurinists might have convinced them that segregation of the sex hybrids does not proceed according to Mendel. At present, the nonscientific character of Mendel's laws can be demonstrated even on *Drosophila*.

At present, we can demonstrate experimentally one of the fundamental theses of the Michurinist genetics—the genetic, qualitative diversity of cells and tissues of plants. It is also possible now to demonstrate experimentally the inheritance of acquired characters (among others, even on *Drosophila*). Michurinists also have experiments that demonstrate convincingly that hybridization is not a recombination of constant genes, as the Mendelists-Morganists happen to think, but a means of obtaining "unstable" organisms, from which, by rearing, it is possible to create new forms of plants, possessing properties absent from either of the parents. Michurinists have at their disposal ample experimental materials which expose the antiscientific character of the chromosome theory of heredity.[10] But our Morganists have no desire to analyze scientifically the facts obtained by the Michurinists.

The Morganists-Mendelists often cite the tetraploid kok-saghyz as proof of the practical achievements obtained on the basis of the chromosomal theory. Whereas to us it seems that this fact has no bearing on the chromosomal theory. The tetraploid kok-saghyz is obtained by means of external factors acting on the organism, whereas the "soul" of the chromosomal theory lies precisely in this: the development of organisms and their mutability are predetermined by unknowable forces hidden in the chromosomes. Our Morganists often recall the tetraploid kok-saghyz.

Why do they do this? Here we must make a point that Academician T. D. Lysenko does not belong to the type of person who rejects anything practically useful. Quite the contrary, he always welcomes everything useful for our work. And he has studied the tetraploid kok-saghyz far better than it was studied by the Morganists. Morganists need this kok-saghyz for the sake of rendering easier their defense of their antiscientific positions in biology. They raise a great to-do about it with a single purpose —the purpose of legalizing the ideas of Mendelism-Morganism in our biology. But it is doubtful that this approach will bring them victory. No subterfuges have the power to save the sterile,

[10] It would be interesting to check these data if they were ever published in a precise form.

antiscientific, chromosomal theory from a complete exposure. Within the last few years, the Mendelists-Morganists, in defending their ideas, had begun to resort frequently to even less admissible methods of scientific struggle. They began to masquerade as Michurinists, and to say that they are for Michurin, and that if they are fighting, it is only against Academician Lysenko.

Academician B. M. Zavadovski went even further. He said that he was practically the first to stand up and fight against Morganism-Mendelism.

We remember well what kind of a fight against Mendelism-Morganism it was on the part of B. M. Zavadovski. It was a fight according to the rules expressed so well in the popular proverb: "When lovers fight, they are merely amusing themselves." The best evidence is his speech at the present session. B. M. Zavadovski, simultaneously with his declaration of his fight against Mendelism-Morganism, has, in effect, expounded here the views of such "opponents" of Morganism as the hundred-per-cent Morganists, Professor Dubinin, and his closest co-worker, Romashov.

This is revealed best of all in the attitude toward Michurin.

A few years ago, Romashov, distorting the facts, wrote that in all works of Michurin there exist no data which "would contradict the basic principles of contemporary genetics and of the chromosomal theory of heredity."

The differences of opinion between Michurin and Mendelists-Morganists, according to Romashov, come down merely "to the specificity of the objects with which Michurin was working." There is no need to say that this is falsification of Michurin, pure and simple.

Academician T. D. Lysenko has shown that the laws, established by Michurin for fruit trees, are real and applicable to the entire plant kingdom. He has discovered the panbiological significance of the Michurin theory. Perhaps, this is precisely why the Morganists are now striving at all costs to drive a wedge between Michurin and Lysenko.

Romashov, guided by Dubinin, falsifies Michurin. But then he is, after all, a hundred-per-cent Morganist. Whereas B. M. Zavadovski said, in effect, the same things as Romashov about Michurin's attitude toward Mendelism, yet he had asserted simultaneously that he is an opponent of Mendelism, that he has fought the Morganists.

B. M. Zavadovski said that Michurin ought to be read in the original, obviously hinting that Lysenko distorts Michurin.

Here it is necessary to say that there are two methods of reading any work. In this case, we are dealing with the works of Michurin. One method may be designated as the Lysenko method, the other as the Zavadovski-Morgan method. The Lysenko method consists in a daily reading of Michurin's works and in finding there the ways to solve the actual problems of our contemporary theory and practice, a reading of Michurin with the purpose of incessantly developing and perfecting him. This is a creative method of studying Michurin.

The B. M. Zavadovski method, to put it mildly, is a scholastic method. He apparently has to read Michurin's work only in order to ferret out of them a confirmation of his own, long-since-fixed and already ossified, formalistic ideas.

Doctor of Biological Sciences Rapoport, who had participated in the debates, tried to impress upon his listeners that, in the future, Morganists will bring happiness to mankind with their great discoveries. Let us note that this idea does not belong to him exclusively—it is the idea of all our home-grown Morganists. All of them are trying to prove that in the past Mendelism has given a great deal of what was useful in practice. It will give even more in the future. Our contemporary Morganists are straining with all their strength to write on their credit side the strains produced by Lisitzyn, Shekhurdin, Yuryev, and other well-known selectionists. Let us admit for the moment that, in the past, Mendelism-Morganism participated in the creation of some now-widespread strains. Then, at once, arises the question: Why has Mendelism-Morganism become impotent today, why is it that selectionists do not use it today to establish new strains? It appears that Mendelism-Morganism was fruitful in the past, will be fruitful in the future, but today it is sterile. It appears that Mendelism so far did not furnish us with a new method and is asking us to wait, while the old methods have become obsolete and are not used. Such a "state," as a rule, is not characteristic of either life or truth.

As for the strains now appropriated by the Mendelists-Morganists, so far as their origin is concerned, Professor Zhegalov once made the right comment. He wrote that these strains could have been obtained only by way of selection, while the latter could have been conducted only on the basis of a firmly estabished fact of the existence of numerous small forms among all self-pollinators, including wheat as well. The analytical selection method, according to which these strains were established, concludes Zhegalov, allows us to understand an aphorism attributed to Jordan: "In order to obtain a new strain, it is first necessary to have it." Our best selectionist strains are the result of selection from the local peasant strains. Our country gives their due to the selectionists who have performed this selection. But Mendelism-Morganism has played no part whatever in their work.

The Morganists have tried to suppress the development of the Michurin teaching. They have kept back all the young scientific workers who formerly occupied the positions of the chromosomal theory of heredity, but who, under the pressure of experimentally produced facts, came to an agreement with this or that principle of the Michurin doctrine. Professor N. P. Dubinin is particularly outstanding in this respect.

One can often hear from the Morganists that it is impossible to have a discussion with Academician Lysenko, that he hushes them up. The truth lies not in this, but in the fact that in our Country the Michurinist trend in agrobiology has been, and still is, in the way of the Morganists. In the light of the efficacy of the Michurin doctrine, the sterility of Morganism becomes all too obvious, and this is why the Morganists are shouting about suppression.

With the arrival of Academician Lysenko at the Lenin All-Union Academy of Agricultural Sciences, they began there to put into practice one of the great principles of science.

K. A. Timiryazev once expressed the essence of these principles in the following words: Work for science, write for the people. T. D. Lysenko consistently carries out this scientific principle. But he supplements it with a still more efficacious, Leninist principle. At one time Lenin made a remark that a concrete analysis of a concrete situation is the very soul of the dialectic.

With T. D. Lysenko, with the Michurinists, all research is subject to the task of solution of this or that important practical problem. On this foundation, the Michurin doctrine grows and gets stronger.

Living deeds are enemies of formalism. In the light of the living, Michurinist deeds, which are growing stronger and stronger in our country, the scholasticism, metaphysics, and sterility of Morganism have become particularly obvious. And this is what turns out to be "repression" for the Morganists. They do not want to get busy with living deeds, that would cure them of formalism.

Science is a living organism that develops truth, as Hertzen used to say. The Soviet science is all the more a living organism, because it is a science of the people. And this living, healthy organism will know how to free itself from the deadening, reactionary Weismannism.

This is attested by the present session of the Academy that bears the name of the immortal Lenin and is guarded by the paternal care of the great Stalin. *(Applause)*

10

Academician I. I. PREZENT

The present session is devoted to surveying the achievements and the future course of the science of biology. The summing up of trends in biology, performed at this session, is not to be measured merely by a brief time period. Essentially, here was posed and, I am certain, solved the question of the larger trends in biology for many decades to come.

At present, a great divide has been delineated between the Mendel-Morgan (Weismann) trend, on one hand, and its opposite, Michurin trend, on the other. In this connection, it is extremely important to examine the attempts, made both at this session and outside, to find some common ground for a reconciliation of these two opposed trends.

Is it at all possible?

It must be admitted that such a line of "reconciliation" is perhaps possible; but to achieve it, it is necessary for the Morganists and their sympathizers, as well as those who are trying somehow to acquit them "statistically" *(Laughter in the audience)*, to give up their position that there exist two histories, of which one, the phylogenetic one—i.e., evolution of species, races, and varieties—is independent of the other—i.e., the history of the individual development of an organism—and that the latter in no way determines the former. They must give up their assertions that development of races and varieties is in no way determined by the special mode of life and the special conditions of development of the individual. They must give up the false notion of a special "hereditary substance" which is an entity in its own right, independent of the body, in relation to which the rest of the body is of no importance, and on which the body cannot exercise any specific effect. They must give up their position that a gamete (a sex cell) is pure and preserves its pristine purity untouched by the influences deriving from the body and the conditions of its life.

Briefly speaking, in order to "reconcile" Morganists with Michurin's teachings, the Morganists must give up every single theoretical position of their false doctrine. Michurin's science, Michurin's biology, will not let itself be lured in any other direction. Nor will success crown those who are trying to falsify the very Michurin doctrine by making this progressive teaching appear similar to the reactionary Morganism, announcing afterward that they too are going in "Michurin's direction."

At the moment, in our country there are not many open and avowed Morganists. For that, indeed, they would all have to be Dubinins. *(Applause)* [This is apparently a pun. "Dubinin" is the name of a geneticist; *dubina* is a contemptuous term, equivalent to "blockhead."]

First of all, I shall try to analyze these attempts to turn Michurin's doctrine into an imitation of Morganism—attempts undertaken, for instance, by Alikhanyan. The method he uses is not very original and is as follows: Michurin had fought against Grell. Grell stood for the influence of the stock on the scion. And Michurin was against Grell. Consequently, concludes Alikhanyan, Michurin was against the effect of the stock on the scion, against vegetative hybridization.

Esteemed "Neo-Michurinist," Comrade Alikhanyan! You indulge here in an elementary logical fallacy, a fallacy too easy to perceive, for you to be able to get away with it. You declare that since I. V. Michurin was *against Grell's* theory of the influence of the stock on the graft, it means that he was *in general* against *any* effect of the stock on the graft. This kind of "logic" will not be permitted to pass through the gates of Michurin's doctrine.

Anyone in the least familiar with Michurin's works knows that, having discovered the causes of Grell's failures, Michurin himself had discovered the *law* of the mutual interaction between the stock and the scion, thus simultaneously laying down the foundation of the theory of plant development. As a result, Michurin had arrived at a conclusion, in which he had summed up his numerous works on vegetative hybridization: "I consider the question of vegetative hybrids as definitely proved."

Needless to say, Michurin's doctrine has found further rich development in the works of a large and goodly group of Michurinists who are proud to be headed by T. D. Lysenko—a master of experimentation, a mind as sharp as a surgeon's knife, one of our subtlest thinkers.[11] In a short period of time, this movement has posed and solved many problems of biological and bioagronomic sciences, both in their most general and theoretical aspects and their most concrete form, thus making no small contribution to our national economy. It is impossible to enumerate all these problems—the ones already solved and those that are in the process of solution. However, I shall allow myself to pause on one of them, namely, the problem of vegetative hybridization.

[11] This statement deserves comment. Prezent, himself, is one of the most intelligent of Lysenko's followers and is widely credited, outside of Russia, with having supplied Lysenko with ideas and with having steered him to success. He must be aware of Lysenko's stupidity and scientific limitations. The question naturally arises: Is Prezent's eulogy a secret jest at Lysenko's expense? Or could Prezent be, in his own mind, describing himself as the clever fellow who has been using Lysenko as a figurehead?

Academician Zhukovski makes a remark: what if the effect of vegetative hybridization is actually due to inadvertent pollination by an alien variety? Let me assure you that Michurinists are far more experienced experimenters than the Morganists, and that the possibility of such an elementary error had certainly been foreseen and eliminated. Assumption of alien pollination, in order to deny the existence of vegetative hybrids, has as much weight as the pronouncements of the hostile, foreign biologists who declare that all Lysenko's works are worthless, since "it is well known that Russian strains are not pure." This is particularly the way Hudson and Richens write in their big survey, *New Genetics in the Soviet Union.* But you will agree that if one leads a good life, he would not argue from the pure to the impure. *(Laughter, prolonged applause)* [This is both a pun and a literary allusion. "Impure" is one of the Devil's nicknames; "the good life" is an allusion to a well-known humorous story by Gorbunov-Posadov.]

The Morganists spin many a yarn about the Michurinists. Thus, Academician Zhukovski mentioned in his speech that he has heard from someone that Lysenko and his followers deny that a Russian scientist, S. G. Navashin, discovered double fertilization, which leads to the formation of endosperm.

Now let me ask: who is continuing the progressive aspect of Navashin's works? What have Morganists done to explain the phenomenon and the effect of double fertilization theoretically from the Darwinian standpoint? Nothing! Whereas we humble Michurinists have taken up this problem and are working it out from the standpoint of the general Darwin-Michurin idea about the usefulness of fertilization in general, and of cross-fertilization in particular. Let me here point out briefly that, after working out this problem experimentally, I have come to a conclusion that double fertilization leads to a formation of hybrid nutriment, comprising a wealth of species adaptation. This is something contrary to the forced self-pollination. When in our experiments we raised plants from germs isolated from the endosperm, we obtained phenomena analogous to inbreeding: plants of stable, straight-line strains would turn out to be highly diversified, and would change characters, up to varietal ones.

Esteemed comrades! Our Mendelist-Morganists are now trying to pass for Darwinists, sometimes calling themselves the true, or else, orthodox Darwinists. And, of course, above anyone else, Academician Schmalhausen is proclaimed to be such a Darwinist.

To analyze Schmalhausen's errors in full would require a line-by-line quotation of his works. This possibility is not at my disposal, and therefore I shall pause only before a few of his erroneous assumptions.

First of all, let me point at the following paradox. Schmalhausen declares himself to be an enemy of Lamarck. A well-known, bourgeois biologist-idealist, Cope,[12] represents one of the strongest pillars of idealism in biology, and takes from the Lamarckian doctrine its most rotten, most backward elements, while discarding its progressive and materialistic aspects. As is well known, Cope has published his so-called "doctrine of nonspecialization." The essence of this doctrine is that every neoformation must have as its source a nonspecialized organization, and the farther the process of specialization of an organism has progressed, the fewer are the chances of any kind of neoformation. From this "doctrine," Cope's followers have made some appropriate deductions. They ask, for instance: could man have descended from some ancestor of our contemporary ape? And, after scrutinizing all the fossil ancestors of the ape, they argue that every one of them has certain marks of specialization. And true enough, seas and continents were never populated by mere schemata, but have been always inhabited by adapted—and thus, to some degree, specialized—organisms. But if the "doctrine of nonspecialization" is correct, then none of the fossil apes could be considered as man's ancestor, and man has a source of origin of his own, independent from all the rest of animals.[13] It is easy to see that this type of argument is thoroughly anti-Darwinian and would openly lead to popery.

Our leading Russian scientists, such as Sushkin, have refuted this false, antiscientific "doctrine" of Cope. Whereas Schmalhausen, in spite of all its absurdity, accepts it. More, he actually eulogizes Cope in his book *The Problems of Darwinism,* which he has recommended from this tribune as a thoroughly Michurinist and Darwinist book. Here is what he writes: "We must pay attention to the data of paleontology, already noted by Cope and confirmed many times since. . . . New forms thus always originate from the less-specialized ancestors—representatives of the preceding epoch. . . . These deductions follow from the sum-total of our knowledge."

[12] E. D. Cope (1840-1897). The Russians devote much of their attention to nineteenth rather than to twentieth century biology.

[13] This is one of the crudest *non sequitur's* of the entire conference. If the fossil apes are too highly specialized in a nonhuman direction to be our ancestors, it may only mean that the fossils of our ancestors, contemporary with the fossil apes, have not yet been found. There is nothing here to indicate that the fossil apes and ourselves are not descended from a common ancestor who lived in a still earlier period. Certainly some of the very old fossils recently discovered in South Africa are very close to the human line. Prezent could not have been unaware of his *non sequitur.* That he used it at all is an indication that he had a secret contempt for his audience. That his audience received his address enthusiastically perhaps indicates that the contempt was deserved.

I shall take the liberty to ask Academician Schmalhausen whether, in his belief in Cope's doctrine, he would, for instance, consider that our contemporary penguins, whose pirn [quill?] is less specialized than that of the primal bird *(Archaeopteryx)*, could not descend from the latter, but must have a separate ancestor, apart from all the other birds? How will Schmalhausen explain the fact that evolution is still going on among the present plants and animals, which are already sufficiently specialized?[14] Or perhaps he subscribes to the End-of-Evolution theory, propounded by Julian Huxley, according to whom evolution is now hanging by a single, thin thread—the thread of human evolution alone. May I ask how can Cope's delirium be considered a valuable contribution to Darwinism? May I ask how can such an anti-Darwinian concept be presented in a book which bears the title *The Problems of Darwinism* and is recommended by the Ministry of Higher Education as a scientific textbook for our universities?

Schmalhausen claims that all these errors have been attributed to him falsely. Then let me ask him—does he still subscribe to the following of his former statements:

> Within the boundaries of pure lines, selection is powerless. Due to a misunderstanding, these facts have been interpreted as contradicting the theory of natural selection, although Darwin himself had stressed that, of course, natural selection does not create new changes, but merely accumulates them. In nature, there are no pure lines, and selection always operates among more or less heterogeneous populations, which possess a tremendous range of all kinds of individual characters.

It is easy to see that this Morganist assertion—that selection is ineffective in "pure lines"—is regarded by Schmalhausen as inapplicable to wild nature merely because here there are no "pure lines." Yet he considers the above assertion entirely applicable and correct where such "pure lines" do exist—i.e., in the realm of selective breeding. One wonders where did Academician Schmalhausen spend these last few years, and in what spheres dwelt his thoughts, if he is so completely ignorant of the facts of selection, if he does not know that thousands of people have already confirmed T. D. Lysenko's thesis of mutability of pure lines.[15]

[14] Prezent obscures the question by the undefined term "*sufficiently* specialized." Evolution generally proceeds toward *greater* specialization and from *less* specialization.

[15] Lysenko cannot honestly claim this "thesis." Evolution in a pure line of *Difflugia* was proven by H. S. Jennings in 1916. See *Genetics*, Volume I. The question of pure lines and evolution are discussed adequately in practically all textbooks of genetics.

In his speech, Academician Schmalhausen has recommended himself as a continuator of the works of Severtsov's, and claims that he had been thus ordained by Severtsov himself. I am in no position to contradict this. But if Schmalhausen was indeed initiated by Severtsov, then it must be admitted that the initiate has failed to justify the highly honored rank conferred upon him. We are fully entitled to say that, under the guise of "continuing" Severtsov's work, Academician Schmalhausen has merely multiplied and classified words; that while he pretends to develop Severtsov's teachings, in reality he is merely cluttering them up with "allomorphoses," "telomorphoses," "catamorphoses," "hypomorphoses," "hypermorphoses," and up to a certain "epimorphosis," which implies that human history is placed in the general order of classification of animal evolution.

As for the fundamental principle in Schmalhausen's pseudo-scientific construction, it is Weismann's autonomization of the organism, which was given a distinct expression by one of the followers of Schmalhausen—Professor Paramonov—who states that "an organism represents an independent system, while its surrounding environment is another system. . . . Trends of environmental changes and those of the mutability of organisms do not depend on each other."

Our Morganists, retreating all along the lines under the pressure of the Michurinist facts, are trying to hold out on the least attacked front. This front is cytology. However, at present, even in the domain of cytology there are available some striking facts, which indicate that the so-called cytogenetics is built on sand.

A Soviet scientist, Professor Makarov, has demonstrated that the so-called uninterrupted continuity of the chromosomes is a myth. Discoveries of a prominent cytologist, Jeffrey, has even forced the editors of such a journal as *Science* to admit that this research appears "to render necessary a revision of our views regarding the relation of the nucleus and its derivative chromosomes to heredity and the determination of sex."[16] The Vice-President of the American Association for the Advancement of Science, and Chairman of its Section of Zoölogical Sciences, F. Schrader, was forced to admit that "in the cytology of *Drosophila* there is a great deal that does not correspond to processes which we regard as standard," and that "most of the foundations on

[16] Prezent is here referring to E. C. Jeffrey, a plant morphologist, Professor Emeritus at Harvard since 1933. The paper in question is "The Nucleus in Relation to Heredity and Sex," *Science* 106:305-8. 1947. The quotation is from a footnote on page 305. This paper was published when *Science* had no editor. No member of the Publications Committee was a cytologist. It would be interesting to identify the author of the footnote.

which present cytogenetics is built require a revision," and that "almost all the cytologists, except for Darlington and his followers, are by now convinced that his factual data are incorrect."

Hugo Iltis, the same Iltis who, when a memorial was being erected to Mendel, used to worship at Mendel's remains, anticipating the recommendation of Academician Zhukovski, now declares sadly:

> These are the black days for the genes. It's disheartening, when your very existence is in doubt. . . . A not inconsiderable slur is cast upon the dignity of the gene—whose chief pride lay in its crystalline purity and constancy—by an accusation of lability and of stability hardly greater than the stability of a lump of sugar dissolving in a cup of coffee.

Mendelism-Morganism has already fully revealed its yawning void; it is, moreover, rotting from within, and nothing can save it.

Unfortunately, the corrupting influence of Morganists has spread even among nonbiologists.

It is time to put an end to this corrupting influence of Morganists on workers in other fields, particularly on philosophers! Philosophers are under an obligation to have their own—and what's more, a correct—point of view as to who has solved the problem of controlling hereditary variations: Morgan and the Morganists, or Michurin and the Michurinists. Up to now, many philosophers have hesitated over these problems. But, after all, there must be a certain limit to hesitations. You cannot be a pendulum when it comes to problems of science. *(Laughter)* The time has come to bring to light the philosophic depths of Michurin's doctrine, and I have faith in our philosophers—they will do it. *(Applause)*

Some of the Morganists who have appeared here have assured us that they too are Michurinists. For instance, B. M. Zavadovski. Trying to prove that he too is a Michurinist, and that he merely wants to unite and reconcile something or other, he said: try to find another such museum as the one of which I am a director; where will you find such an ample demonstration of Michurin's teachings as in my museum?

In this connection, it is interesting to relate the following curious incident. Shortly before the war, I was appointed a member of the commission that was checking upon the activities of this particular museum. And we have discovered that on one side of exhibition boards were mounted exhibits propagandizing the Michurinist trends, while on the reverse side of these movable boards were mounted exhibits propagandizing Morgan's views. So that, depending on the composition of the visiting group, it was possible to turn these exhibits any way you choose.[17] *(Laughter)* This convenient technical method, it seems to me, is but a technical formulation of the ideological principles and concepts of B. M. Zavadovski. *(Laughter, applause)*

If we now turn to the statements of another "semi-Morganist" who had appeared on this rostrum—I. M. Polyakov,—we shall be forced to note the remarkable poverty and lack of content marking his statements. For instance, he says that he cannot separate Michurinists from Morganists who, after all, also influence the organism by various substances, such as X-rays, etc. But there's influence and "influence." There is influence that pervades the entire process of development of the organism, taking into account its history, and there is "influence" of the type of a single blow, with no reference to the biological peculiarities and the history of the organism. Even if this type of blow is softened, still, if it does not follow through the entire life regimen, through its entire development, then such effect can be but accidental. This is not the right road for planned selection.

Speaking of his disagreements with T. D. Lysenko, I. M. Polyakov has pointed out that it is not entirely clear to him—how does Lysenko solve the problem of purposiveness? And yet in all the works of T. D. Lysenko and of other Michurinists, there is traced a distinct line separating the needs of organisms under certain definite conditions from the purposiveness of their deviations. It is pointed out that selection, and selection alone, renders a developing need purposeful and decides whether it will be passed on to following generations, whether it will accumulate and become a property of the propagating individuals or not.

However, the current Michurin doctrine cannot be limited by the so-called classical Darwinism. The thing to do is not merely to purge Darwinism of its sins, but also to raise Darwinism, to lift it to a fundamentally new level, the level of Michurin's doctrine. Darwinism today is not what it used to be in Darwin's time. The law of selection, in the light of Michurin's doctrine, is not formulated the way Darwin himself had formulated it. This law of selection must necessarily include the role played by the conditions of rearing, and if we are dealing with artificial selection, it emerges as a systematically planned, fostering effect. This level and form of selection were unknown to Darwin and his followers; they were unknown and could not have been known,

[17] This is a most damning admission for an intelligent man to make. Prezent could hardly have been unaware of what he was doing. His jeer gives a real picture of how scientists have to work in a totalitarian state and what expedients they have to adopt to keep their work going. In exposing these conditions, Prezent has performed a valuable service to world science.

but Professor Polyakov ought to know them. After all, it is his duty to read the theoretical works of Michurinists!

Now anti-Michurinists have but one tactical approach left to them—the approach which Timiryazev had once ascribed to the helplessly rabid anti-Darwinists, and which is now fully applicable to the helplessly rabid anti-Michurinists. Different creatures have recourse to different means of defense: a lion defends himself with claws; a bull, with horns; a hare is saved by its fleet feet, a mouse hides in its hole; a cuttlefish muddies up the surrounding water and under the cover of darkness slips away from its enemy. And precisely these tactics of a cuttlefish are invariably imitated by our anti-Michurinists, with this difference only—a cuttlefish, of course, is glad merely to escape from its enemy, whereas our anti-Michurinists, out of their darkness, shower abuse on their adversaries and shout smugly: "We've defeated them! We have won! We've destroyed them!"[18]

These cuttlefish tactics were used, in particular, by Rapoport, during the present session. He declared that the basic concept of Michurin's genetics—the organism's requirement of certain conditions of life—is nothing but Machism. With equal success, Rapoport could have declared that Darwinism, as a whole, is nothing but Machism. After all, the basic concept of Darwinism is adaptation, and Mach, as is well known, has built his false philosophy of "empirio-criticism" on an irrelevant application of this concept to epistemological processes.

Apart from the tactics of a noisy cuttlefish, the Morganists, for lack of scientific argument, employ the tactics of organizational discrimination against Michurinists, whenever Morganists happen to be at the helm of faculties, academies, etc. Professor Beletski has told us from this rostrum how it is done at the Moscow University. The same thing, though probably in an even more inadmissible form, will be found in our Leningrad University as well. The Dean of the Faculty, Lobashov, the Pro-Rector Polyanski, and the group of their adherents, have tried by all possible, and more often impossible, means to banish Michurinists from the faculty domain.

The principles of this little clique, which rules the Department of Biology of the Leningrad University, have been stated recently by Professor Polyanski:

> Without doubt, during the last few years, T. D. Lysenko, in a whole series of his works, has been defending some profoundly erroneous, harmful, anti-Darwinist positions. . . . Erroneous and highly harmful is T. D.

[18] The above paragraph is a very striking sample of the tone of scientific controversy under Communism.

Lysenko's nihilistic denial of all the principles established by genetics, a denial of all the positions of Mendelism-Morganism. . . . Unfortunately, we must state that recently in our Soviet biology there has appeared and is being defended a series of positions highly injurious to Darwinism, highly injurious to the dialectical method itself, as it is reflected concretely in biology. . . . These errors are being aggravated and will become manifest in practical matters. If we follow the road of crude Lamarckian principles, it would mean an incorrect orientation of selective breeding, it may cause the greatest damage to our socialist economy. Let's not shut our eyes and let's not mince words.

Comrades, we may gladly state that our Soviet biologists, armed with Michurin's doctrine, have already smashed Morganism. Let no one be confused by the false analogies of the Morganists about the invisible atom and the invisible gene. A far closer analogy would be that between the invisible gene and the invisible spirit. We are being challenged here to a debate. We shall no longer debate with the Morganists *(applause);* we shall expose them as the representatives of a harmful, ideologically alien, imported-from-abroad tendency. *(Applause)* Michurinists face the future with courage. We have a real leader, and you have —Schmalhausen. *(Violent, prolonged applause)* Morganists are trying to halt the Michurinist movement now by contrasting Michurin with Lysenko, now setting up early Lysenko against later Lysenko, or Lysenko himself against his own adherents. That is exactly how reactionaries would act. For them, every step forward means their own debacle. Morganists want to halt the progressive march of the Michurinist movement. They did not succeed in it, they are not succeeding, they shall not succeed.

The future of biology belongs to Michurin, and to Michurin alone! *(Wild applause)*

IX

The End of the Road

With the morning session of August 7 the meetings came to an end. This, the tenth session, was opened with Lysenko's concluding remarks in which he endeavored to state the consensus of the participants. The spontaneous enthusiasm with which his remarks were received indicates that his summary was accurate and popular. It was, of course, politically expedient to be on his side, but there is no evidence to indicate that the general elation of the members was assumed. The numerous speeches made during the preceding week show conclusively that the overwhelming majority were in agreement with Lysenko and were in his intellectual class. This is worth emphasizing, for an event occurred which might raise a suspicion of the sincerity in those academicians who cheered their president.

The event in question was revealed by Lysenko at the very beginning of his speech. He stated that the Central Committee of the Bolshevik Party had examined his report and had approved it. This naturally ended all opposition. When he finished speaking Zhukovski, Alikhanyan, and Polyakov hastened to recant. They yielded completely and abjectly to the ukase of the political authorities. Nothing like this had happened within the past three hundred years.

Following the recantations a most servile letter was sent by the Academy to Stalin and the meetings came officially to a close, with the adoption of a set of resolutions.

The remainder of this chapter consists of Lysenko's concluding remarks (*Pravda,* August 10), the letter to Stalin (*Pravda,* August 10), and the text of the resolutions (*Pravda,* August 12). The recantations are included in Chapter X.

1

Academician T. D. LYSENKO

Concluding Remarks on the Report on the Situation in Biological Science

Comrades! Before proceeding to the concluding remarks, I consider it my duty to declare the following.

I have been asked in one of the memoranda as to the attitude of the Central Committee concerning my paper. I answer: the Central Committee of the Party has examined my report and approved it. *(Tremendous applause, passing into an ovation. All rise)*

[*Pravda* (August 8) *describes the scene as follows: "This communication by the President aroused general enthusiasm in the members of the session. As if moved by a single impulse, all those present arose from their seats and started a stormy, prolonged ovation in honor of the Central Committee of the Lenin-Stalin Party, in honor of the wise leader and teacher of the Soviet people, the greatest scientist of our era, Comrade Stalin."*]

I proceed now to an account of some of the results of our session.

The supporters of the so-called chromosome theory of heredity who have appeared here denied that they are Weismannists and almost called themselves opponents of Weismann. At the same time it has been clearly shown in my report, and in the many appearances of the representatives of the Michurinist movement, that Weismannism and the chromosome theory of heredity are one and the same thing. Foreign Mendelists-Morganists do not conceal it in the least. In the report I quoted passages from articles by Morgan and Castle, published in 1945. In these articles it is directly stated that the basis of the chromosome theory of heredity is the so-called doctrine of Weismann. Weismannism (and this is idealism in biology) is any idea on heredity recognizing the division of a living body into two principally different substances: the ordinary living body, apparently not possessing heredity but subject to changes and transformations, i.e., to development, and the specific hereditary substance, apparently independent of the living body and not subject to development in relation to the conditions of life of the ordinary body, called a soma. This is indisputable. None of the attempts of the defenders of the chromosome theory of heredity (whether they attended the session or not) to give their theory a materialistic appearance change the essentially idealistic character of this theory. *(Applause)*

The Michurinist movement in biology is materialistic because it does not separate the characteristic of heredity from the living body and the conditions of its life. Without heredity there is no living body, without the living body there is no heredity. The living body and its conditions of life are indissoluble. Should the organism be deprived of its conditions of life, it becomes extinct as a living body. According to the Morganists, heredity is detached and isolated from the mortal living body or, in their terminology, from the soma.[1]

Out of our differences with Weismannism, significant in principle, arises the divergence on the important historical problem of the inheritance of acquired characteristics both by plants and animals. Michurinists proceed from the possibility and necessity of the inheritance of acquired characteristics. Voluminous factual material, demonstrated at the present session by its participants, again completely confirms this position.[2] The Morganists, including those participating in the present session, are unable to understand this position not having entirely broken with their Weismannist notions.

For some it is still not clear that heredity is inherent not only in chromosomes, but also in any particle of a living body. Therefore they want, so to speak, to see with their own eyes an example of a transmission of hereditary properties and characteristics from one generation to another without a transmission of chromosomes.

For these problems, incomprehensible to Morganists, it is best and clearest of all to reply with a demonstration and explanation of our widely conducted experiments on vegetative hybridization. Vegetative hybridization was worked out as early as the time of I. V. Michurin. Experiments on vegetative hybridization show indisputably that everything living, all cells and all body particles, and not only the chromosomes, possess heredity.[3] You know heredity is determined by a specific type of metabolism. If you are able to change the type of metabolism of a living body, you will change the heredity.

Academician P. M. Zhukovski, as is worthy of a Mendelist-

[1] This misstatement is repeated consistently.

[2] In all the fifty-six papers given at the sessions, not a single specific fact was cited to prove the inheritance of acquired characters. Lysenko consistently confuses grandiose claims with scientific proof. He seems to be completely ignorant as to what is demanded of a scientific experiment.

[3] It is statements like this, published without evidence, which show Lysenko's ignorance of scientific method.

Morganist, cannot imagine a transmission of hereditary characteristics without a transmission of chromosomes. He cannot visualize that the ordinary living body possesses heredity. Heredity, according to his view, is possessed allegedly only by chromosomes. Therefore, he cannot see the possibility of producing hybrids in plants by means of grafting; hence he cannot visualize the possibility of the inheritance of acquired characteristics by plants and animals. I have promised Academician Zhukovski to show vegetative hybrids and here, now at this session, I will have the pleasure of showing them.

For instance, in the field of grafting, a potato-leaf variety of tomato was used, i.e., with leaves, not crosscut as is usually the case in tomatoes, but resembling potato leaves. The fruit of this variety are red and elongated.

The other variety of tomato involved in the graft has ordinary leaves, as all are accustomed to see on tomato plants—crosscut; its fruit in mature view is not red but white or yellowish.

The variety with the potato-like leaves is used in the graft as a stock (i.e., the other variety is grafted on it), and the crosscut-leaf variety is used as a scion.

During the year of the grafting no changes were observed whether on the scion or the stock.

The seed from the fruit grown on the scion and from the fruit grown on the stock was gathered. The seed collected was then sown.

From the seed which had been collected from the stock's fruit, plants grew which in the main were not different from the original variety, i.e., with potato-like leaves and elongated red fruit. Six of the plants did not have potato-like but crosscut leaves. Several of these plants had yellow fruit, i.e., both leaves and fruit were changed correspondingly to the influence of the other variety, the former graft.

Academician P. M. Zhukovski has expressed doubt as to the purity of the experiments on vegetative hybridization, indicating that it was possible a cross-pollination of varieties took place here, i.e., sexual hybridization. But try, Comrade Zhukovski, to explain the results of the experiment demonstrated by me.[4]

[4] Experiments of this type have been performed for generations. Claims, such as Lysenko's, have never been substantiated. As we do not know the genetic composition of either his stock or his scion, we cannot accept his interpretation. The described results could be caused either by the Mendelian segregation of a heterogeneous type or by pollen contamination. In no place is Lysenko's intellectual limitation demonstrated so well as in the next five paragraphs, where he lists the facts upon which his conclusions are based. When he asks, "If it were possible for crosscut leaves to appear in plants of the second generation from cross-pollination, then why is the fruit yellow, and not red?" he shows that Mendelian segregation and recombination are a complete mystery to him, even when only two pairs of factors are involved.

It is well known to all having to do with hybridization of tomatoes that in a cross-pollination of crosscut-leaf, yellow-fruit varieties with potato-like leaf, yellow-fruit varieties that in the first generation the leaves must be crosscut, and the fruit certainly red.

And yet what was derived in these experiments? The leaves truly are crosscut but the fruit, you know, is yellow and not red. How is it possible to explain the results described by the usual cross-pollination?

Here is fruit of another of the vegetative hybrid plants mentioned. On this plant the leaves are likewise crosscut and the ripe fruit on the cluster, as you see, are of two kinds—one is red and the other yellow. The phenomenon of diversity within the limits of a plant is in general fairly frequently evident among vegetative hybrids. It must be kept in mind that vegetative hybridization is not the usual method of uniting varieties; it is not the procedure which was developed in the process of the evolution of these plants. That is why as a result of grafting organisms are frequently produced which are unstable, and thus diversified.

During the year in which the graft was made and even in the first seed generation, it is not possible to observe in all plants easily recognizable changes. Regardless of this, we already have grounds to assert that there is no grafting of a developmentally young plant which would not produce a modification of heredity. As evidence of this condition we are continuing to carry on work with vegetative hybrids of tomatoes at the Institute of Genetics of the Academy of Sciences USSR.

I proceed to a demonstration of plants of the second seed generation from the same grafting, but from seed collected from plants which did not give visible changes in the first seed generation. The leaves of a series of plants in the second seed generation proved to be changed—in appearance they were not potato-like, but crosscut, and the fruit was not red but yellow. In this case likewise there is no basis to doubt the purity of the work or to speak of the possibility of cross-pollination. As you know, these plants had potato-like leaves and red fruit in the first generation. If it were possible for crosscut leaves to appear in plants of the second generation from cross-pollination, then why is the fruit yellow, and not red?

Thus, we see, that as a result of graftings there are produced directed, satisfactory changes; there are produced plants combining the characteristics of varieties united in grafting, i.e., real hybrids. New growths are likewise observed. For instance, here in the offspring of the same grafting are plants bearing small fruit as in uncultivated forms. But everyone knows that new growths are observable even with sexual hybridization in spite of the transmission of characteristics of the parental forms to the offspring.

It is possible to add many more examples of the production of vegetative hybrids. Without any exaggeration we have hundreds and thousands of them in our country. Michurinists not only understand how vegetative hybrids are produced, but produce them in great quantity on the most diverse crops.

I have dwelt on vegetative hybrids as educational material having unusual illustrative significance. As you know, not only Mendelists, but likewise certain materialists not having seen vegetative hybrids may not believe that any living thing, any part of a living body, possesses heredity in the same way as chromosomes. It is easy to demonstrate the condition mentioned in examples of vegetative hybridization. As you know, chromosomes cannot pass from a stock to a scion and vice versa. No one disputes this. In addition such hereditary characteristics as fruit color, fruit shape, leaf-shape, and others are transmitted from the scion to the stock and conversely. Show us then the sort of characteristics, even in tomatoes, which it would be possible to unite from two varieties into one by means of sexual hybridization and which it would be impossible to unite and which have not been united by Michurinists by means of vegetative hybridization.

Thus, experiments on vegetative hybridization indicate faultlessly that any part of a living body, even plastic substances and saps which are exchanged by the scion and stock, possess hereditary qualities.

Is the role of the chromosomes belittled by what has been said? Not in the least. Is heredity transmitted through chromosomes during the sexual process? Of course, how could it be otherwise!

We recognize chromosomes and do not deny their presence. But we do not recognize the chromosome *theory* of heredity; we do not recognize Mendelism-Morganism.[5]

I remind the participants of the session: Academician P. M. Zhukovski promised that if I show him vegetative hybrids, he will believe and will revise his attitude. I have now fulfilled my promise to show vegetative hybrids. But I must remark, first of all, that it has been possible to see such hybrids by the tens and hundreds in our country for more than ten years; and in the second place, is it possible that Academician Zhukovski, a botanist, does not know what is well known, if not to everyone at least to numerous gardeners, namely, that much has been and is being done in ornamental gardening with the idea of changing the heredity of plants by means of grafting?

Some of the Morganists appearing at the session have stated

[5] See page 9.

that, together with the chromosome theory of heredity, Lysenko and his followers also allegedly reject completely all experimental facts achieved by Mendelist-Morganist science. Such statements are not true. We do not reject any experimental facts, including facts regarding the chromosomes.

As to the assertion that the Michurinist movement denies the effect on plants of so-called mutagenic factors—X-rays, colchicine, and others—how can this statement be made? We Michurinists can in no way deny the *activity* of these substances. As you know we recognize the activity of the conditions of life on a living body. Then why should we not recognize the activity of such rare factors as X-rays or the stronger poison colchicine and others. We do not deny the effect of the so-called mutagenic substances, but we persistently argue that influences of this type, not penetrating an organism through its development or the process of assimilation and dissimilation, can only in rare instances and then *accidentally* lead to results useful to agriculture. This is not the path of systematic selection; it is not the path of progressive science.

The long and numerous works carried on in the Soviet Union on the production of polyploid plants with the aid of colchicine and other factors similar in effect have in no degree led to those results which were widely claimed by Morganists.

It was repeatedly said and written that the geranium began to produce seed following an increase of the chromosome set. But this geranium did not go into production, and I as a scientist express the hypothesis that it will not, because propagation of the geranium is markedly more practical by slips. As you know it is possible to sow currants from seed but in practice they are propagated by cuttings. The potato can likewise be grown from seed but planting from tubers is more practical. Usually plants which can be propagated both by seed and cuttings (i.e., vegetatively) as a rule are propagated by a vegetative method in production.

This does not mean that we do not regard as progress the fact that the geranium produced is capable of producing seed. If not for production, then for selection work this method may prove useful.

That which has been mentioned about the geranium likewise pertains to mint.

About what sort of polyploids do the Morganists still frequently speak as if about very important achievements? About wheat, millet, buckwheat, and a group of other plants. Yet, according to the declarations of the Morganists themselves, which we have heard here from the platform (for instance, A. R.

Zhebrak), all of these polyploids—wheat, millet, and buckwheat —proved to be for the time being poorly productive as a rule, and the authors themselves are not turning them over to production.

There remains only the tetraploid kok-saghyz. This kok-saghyz is at present being tested in the collective farms in its first year. If it proves to be good, then it becomes clear in itself that it must be introduced into production. As yet, however, according to data of a three-year State variety test, it is not better than the usual diploid varieties, such as Bulgakov's. This year for the first time tetraploid kok-saghyz has been introduced to testing on collective farms. After the passage of two or three years, experience will show how good it is. I sincerely hope that this kok-saghyz proves to be the best of all the forms of kok-saghyz. This will be only to the advantage of production.

At the same time it is possible to forget that among the varieties of cultivated plants are many polyploids with whose origin not only colchicine and all of the "mutagenic" theory, but also in general all the theory of Morganism-Mendelism, has no connection whatsoever. As you know, people did not know for centuries that many good varieties, for instance, of pears, are polyploids. Just as large a number of similarly good varieties of pear that we have in production are not polyploidic. It is possible even out of one of these facts to reach the conclusion that the category of a variety is not determined by the number of chromosomes.

There are good and poor varieties of hard, 28-chromosome wheat and there are good and poor varieties of soft, 42-chromosome wheat.

Is it perhaps not clear that it is necessary to conduct selection not according to the number of chromosomes, not according to polyploidy, but to good variety characteristics and properties?

After the production of a good variety it is possible to determine the chromosome number. To whom would it occur to discard a good variety only because it proved to be a polyploid or a nonpolyploid? No Michurinist, no serious person in general, can thus raise the question.[6]

Our Morganists, including often those of the present session, as evidence that their theory is effective, frequently refer to such

[6] Here is another instance of a characteristic misrepresentation of an opponent's position. No geneticist advocated discarding a good variety because of the number of its chromosomes. Nor do geneticists assume that the more chromosomes an organism has, the better it is, although sometimes a doubling of the chromosomes may produce characters which we find desirable. There is generally another object in altering the chromosome number. Doubling it by artificial means has often made desirable but sterile hybrids fertile. It thus offers a possible source of superior varieties.

widespread varieties of grain, as, for instance, Lyutestsens 062, Melyanopus 069, and some of the other old established varieties, produced so to speak on a basis of Morganism-Mendelism. But, as you know, the development of these varieties has no connection with Mendelism. How, for example, were such varieties as Lyutestsens 062, Melyanopus 069, Ukrainka, and others developed? They were developed by a long-established method of selection from native varieties.

I refer to the words of Professor S. I. Zhegalov. In the work *Introduction of Agricultural Plants into Selection,* he wrote

> . . . under ordinary agricultural conditions one has to deal not with pure forms but with "varieties," representing more or less complex mixtures of different forms. . . . Almost the very first, the Spanish botanist Mariano Lahaska turned attention to this fact in the first quarter of the nineteenth century (long before the appearance of Weismannism—T. L.), publishing his observations in Spanish. There is a very interesting story about how he visited his friend, Colonel Le Couteur, at his estate on the island of Jersey; while making the rounds of the estate with the host, he turned the latter's attention to the marked diversity of plants and suggested the idea of making a selection of individual forms for subsequent cultivation in pure strains. Le Couteur made use of this suggestion, chose 23 different forms from his field, and began to test their comparative qualities. As a result of this testing one of the isolated forms was acknowledged as the very best and in the year 1830 was released for sale under the name of a new variety "Talavera de Bellevue"! Similar work has been carried out many times since and has led to the release of many valuable varieties. The principle is a separation of the initial mixtures into their component parts; that is why such a method of selection obtained the name of *"analytical selection"!* At present this method is basic in work with self-pollinating plants and is systematically applied by all stations, particularly at the beginning of work on plants formerly little affected by selection.[7]

Further, Professor S. I. Zhegalov writes: "The method of analytical selection makes comprehensible the aphorism ascribed to Jordan: 'In order to produce a new variety, it is first necessary to master it.' "[8]

Comrade Shekurdin, was the form of wheat, known now as variety Lyutestsens 062, among the native variety of Poltavka or not? (*Voice from the audience:* It was, definitely.) It is the same story with the forms which are known as varieties Ukrainka or Melyanopus 069.

And that is why S. I. Zhegalov accepts the aphorism that when

[7] S. I. Zhegalov, *Introduction of Agricultural Plants into Selection,* pp. 79-80. 1930.

[8] *Ibid.,* p. 83.

utilizing the method of analytical selection to obtain new varieties it is necessary first of all to master them. The varieties indicated, to which our Mendelists refer, were in fact so obtained.

But we Michurinists cannot agree with Professor S. I. Zhelagov—with such an interpretation of Darwinist selection. It is possible to begin selecting plants, even with weak but useful characteristics barely noted, in order to obtain a development of these useful characteristics by repeated selections under proper cultivation and strengthening of plants. As it is clear to everyone, the Darwinist method of selection described by us has no connection with Mendelist-Morganist theories.[9]

It should be noted that formerly varieties were brought out only on the basis of the method just described, but even now it is used and will be used. It is a useful method. Practical people—selectionists who use this method successfully—should be appreciated and upheld.

We not only do not reject the method of continuously improving selection; but, as it is well known, we have always insisted on it. The Morganists meanwhile ridiculed improving repetitive selections in grain-growing practice.

Weismannism-Morganism never was and never will be a science which would offer the opportunity to create systematically new forms of plants and animals.

It is characteristic that abroad, for example in the United States of America, homeland of Morganism where it is so highly extolled as a theory, this doctrine is not applied in agricultural practice because of its uselessness. The theory of Morganism is considered by itself, while practice goes its own way.[10]

Weismannism-Morganism not only does not reveal the real regularities of animate nature but, being an idealistic doctrine to the core, it leaves an entirely false idea about natural regularities.

Thus, the Weismannist idea on the independence of hereditary characteristics of an organism from the conditions of the surrounding environment has led scientists to the assertion that the property of heredity (i.e., specificity of the organism's nature) is subject only to chance. All the so-called laws of Mendelism-Morganism *are constructed exclusively on the idea of chance.*[11]

In confirmation of the above I will cite examples.

[9] This statement of Lysenko shows his ignorance of the literature and practice of genetics. By applying various types of selection to populations heterozygous for a number of Mendelian factors, it becomes possible to describe natural selection quantitatively.

[10] This statement is false. See U. S. Department of Agriculture *Yearbooks* for 1936 and 1937. Hybrid corn is a creation of Mendelian genetics.

[11] So are the actuarial tables of the life insurance companies. Lysenko here shows his inability to understand mathematics.

"Gene" mutations arise, according to the theory of Mendelism-Morganism, by chance. Chromosome mutations likewise appear by chance. The direction of the mutation process as a result of this is also by chance. Proceeding from these fictitious accidents, Morganists construct their experiments on chance selection of the means of influence on the organism, the so-called muta-genic substances, assuming that by this they will influence their fictitious hereditary substance and hope to obtain by chance that which may be accidentally useful.

According to Morganism, the divergence of the so-called material and paternal chromosomes during the reduction division is likewise subject to pure chance. Fertilization, according to Morganism, does not proceed selectively, but on a principle of chance meeting of the sex cells. Hence, the segregation of characteristics in the hybrid offspring is likewise by chance, and so forth.

In accordance with this type of "science," the development of an organism is not accomplished on the principle of selectivity of conditions of life from the surrounding external environment, but again on the principle of perception of substances received from outside by chance.

In general, animate nature appears to the Morganists as a chaos of chance and torn phenomena outside of essential relations and conformities with the laws of development. Environment is conditioned by chance.

Not being able to reveal the conformity of the laws of development in animate nature, the Morganists have been forced to resort to the theory of probability and, not understanding the concrete content of biological processes, they transform biological science into bare statistics. Not in vain do the foreign statisticians—Galton, Pierson, and now Fisher and Wright—likewise consider themselves founders of Mendelism-Morganism. Probably for this reason Academician Nemchinov declared here that for him as a statistician the chromosome theory of heredity was readily comprehensible. *(Laughter, applause)*[12]

Mendelism-Morganism is built only on chance, and with this alone this "science" contradicts the requisite relationships in animate nature, dooming practice to fruitless expectation. Such science is devoid of effectiveness. Systematic work, objective practice, and scientific foresight are impossible on the basis of such a science.

A science which does not give practice a clear perspective, a power of orientation, and a confidence in attainment of practical

[12] The Michurinists uniformly deride mathematics.

aims does not merit being called a science. (Applause)

Such sciences as physics and chemistry have freed themselves from chance. That is why they became exact sciences.

Animate nature was developed and is developed on a foundation of the most strict and inherent rules. Organisms and species are developed on a foundation of their natural and intrinsic needs.

By getting rid of Mendelism-Morganism-Weismannism from our science we banish chance out of biological science. (Applause)

We must keep in mind clearly that science is the enemy of chance. *(Tremendous applause)* That is why the transformer of nature Ivan Vladimirovich Michurin put forward the motto: "We can not expect favors (i.e., fortunate chances.—T. L.) from nature; our task is to take them from her." *(Applause)*

Knowing the practical sterility of their theory, the Morganists do not even believe in the possibility of existence of an effective biological theory. Not knowing *A* from *B* in Michurinist science, they have been unable up to now to visualize that for the first time in the history of biology a real effective theory has appeared—the Michurinist doctrine. *(Applause)*

Proceeding from the Michurinist doctrine, it is possible to foresee scientifically, and ever more and more to free practical plant growers from chance in their work.

I. V. Michurin himself worked out his theory and his doctrine only through the process of the solution of important practical problems, through the process of developing good varieties. Therefore the Michurinist doctrine is in its spirit inseparable from practice. *(Applause)*

Our collective-farm system and socialistic agriculture have met all the conditions for the flourishing of the Michurinist doctrine. It is necessary to remember Michurin's words: "The history of the agriculture of all times and all peoples has in the person of the collective-farm worker an entirely new being, that of a cultivator who has entered into the struggle with the elements with a wonderful technical equipment and who is attacking nature from the viewpoint of a transformer."[13]

> I see [wrote I. V. Michurin] that the collective-farm system through which the Communist Party has begun to carry out the task of the restoration of the land will lead working humanity to actual power over the forces of nature.
>
> The great future of all our natural science lies in the collective and state farms.[14]

The Michurinist doctrine is inseparable from collective- and

[13] I. V. Michurin, *Works,* I, 477.

[14] I. V. Michurin, *Works,* Vol. I, p. 477.

state-farm practice. It is the finest form of the unity of theory and practice in agricultural science.

It is clear to us that a broad development of the Michurinist movement is impossible without the collective and state farms.

Without the Soviet order I. V. Michurin would have been, as he himself wrote, "an unknown hermit of experimental horticulture in czarist Russia."[15]

The strength of the Michurinist doctrine lies in its close association with the collective and state farms, *in the working out of deep theoretical problems by the method of a practical solution of the important tasks of socialistic agriculture.*

Comrades, the work of our session comes to a close. This session is clear evidence of the strength and power of the Michurinist doctrine. Many hundreds of representatives of biological and agricultural science took part in the session's work.

Gathered here from all the corners of our vast country, they accepted an active role in considering the situation in biological science and, convinced by their practical work of many years of the accuracy of the Michurinist doctrine, they wholeheartedly support this movement in biological science.

The present session has shown *the complete triumph of the Michurinist movement over Morganism-Mendelism. (Applause)*

The present session is in truth a historical landmark in the development of biological science. *(Applause)*

I believe that I do not err in saying that this session is a great festival for all the workers of biological and agricultural science. *(Applause)*

Fatherly interest is shown by the Party and the Government for the strengthening and development of the Michurinist movement in our science, and for the elimination of all impediments on the road to its further flourishing. This obliges us to develop the work for the fulfillment of the order of the Soviet people in equipping even more widely and deeply collective and state farms with advanced scientific theory.

We must earnestly place science and theory at the service of the people to increase the fertility of the fields and productivity of animals and to increase the productive efficiency of collective and state farms at an even more rapid tempo.

I call upon all academicians, scientific workers, agronomists, and zoötechnicians to exert all their efforts in close unity with the leaders of socialistic agriculture for the fulfillment of these great and noble tasks. *(Applause)*

Progressive biological science is indebted to humanity's gen-

[15] I. V. Michurin, *Works*, Vol. IV, p. 116.

iuses—*Lenin* and *Stalin—that like a golden fund the doctrine of I. V. Michurin has been added to the treasury of our knowledge, to science. (Applause)*

Long live Michurin's doctrine, the doctrine of the transformation of animate nature to the benefit of the Soviet people! *(Applause)*

Long live the party of Lenin-Stalin, which revealed Michurin to the world *(applause)* and which created in our country all the conditions for the flourishing of advanced materialistic biology. *(Applause)*

Glory to the great friend and coryphaeus of science, to our leader and teacher Comrade Stalin!

(All rise and applaud for a long time)

2

V. I. LENIN ALL-UNION ACADEMY OF AGRICULTURAL SCIENCES

Letter to Stalin

(Read by I. D. Kolesnik)

To Comrade I. V. Stalin
Beloved Iosif Vissarionovich!

The participants at the Session of the V. I. Lenin All-Union Academy of Agricultural Sciences: academicians, agronomists, animal breeders, biologists, mechanizers, and organizers of socialistic agricultural production send you their cordial Bolshevist regards and very best wishes.

Each day and hour the scientists and practical workers of agriculture feel the manifold anxiety of the Communist Party and the Soviet Government for agricultural science and your constant personal participation in the concern of its further development and prosperity.

To you, great creator of Communism, our Fatherland's science is indebted in that through your highly gifted efforts you have enriched and elevated it before all the world, you protect it from the danger of estrangement from the interests of the people, you help it to gain victories over reactionary teachings hostile to the people, you look after the uninterrupted growth of the workers of science.

Continuing the work of V. I. Lenin, you have saved for progressive materialistic biology the teaching of the great transformer of nature, I. V. Michurin, and have elevated the Michurinist movement in biology before the eyes of all science as the only true progressive movement in all branches of biological science. By this the natural-scientific bases of the Marxist-Leninist world outlook, whose all-conquering power is confirmed by all the experience of history, were strengthened even more.

You, our beloved leader and teacher, daily help Soviet scientists in the development of our progressive materialistic science, serving the nation in all its efforts and achievements, a science expressing the world outlook and noble aims of man of the new socialistic society.

The collective-farm system, established under your wise leadership, has disclosed the boundless possibilities for a tremendous raising of the productive forces of all branches of agriculture and has shown its invincible power. The party of Lenin-Stalin has trained amongst the collective-farm peasantry remarkable fighters for high productivity of agricultural crops and in animal husbandry. Michurinist agricultural science, called upon by you to develop more boldly and decisively the scientific investigations on the active transformation of the nature of plants and animals, equips practical men in their struggle for a high level of cultivation of socialistic agriculture. In their turn the progressive people of the collective-farm villages, innovators of agricultural production on the basis of the socialist competition of all the people, enrich our science with new methods and new achievements.

We assure you, beloved Iosif Vissarionovich, that we will apply all our efforts in order to help the collective and state farms to attain an even greater productivity in our socialistic fields and animal husbandry for securing the abundance of products in our country as one of the most important conditions of the transition from Socialism to Communism. We see the possibilities for attaining this great aim through a close unity of science with the nation, with the progressive people of the collective-farm villages—which you always have taught and teach us, Party and non-Party Bolshevists. A science that is fenced off from the people, from practice, is not a science.

Our agrobiological science, developed by the works of Timiryazev, Michurin, Williams, and Lysenko, is the most advanced agricultural science in the world. It is not only the legitimate successor of the progressive ideas of the advanced thinkers in all the history of mankind, but in itself represents a new, much higher level of development of human knowledge about highly advanced agricultural cultivation. The Michurinist doctrine is a new and higher stage in the development of materialistic biology. Michurinist biological science will henceforth likewise develop Darwinism creatively; will steadfastly and resolutely expose reactionary-idealistic, Weismannist-Morganist scholasticism, severed from practical work; will fight against servility before bourgeois science as unworthy of a Soviet scientist; and will liberate investigators from survivals of idealistic, metaphysical ideas. A progressive biological science rejects and exposes the vicious idea as to the impossibility of controlling the nature of organism by means of living conditions of plants, animals, and microörganisms under the control of man.

Science must teach investigators to maintain their search of ways and means for the control of nature for human needs.

We are inspired along this path in science and practice by the triumphant teaching of Marx-Engels-Lenin-Stalin.

Along this path we are inspired by your instructions about progressive science as serving the people, valuing tradition, but not afraid to take up arms against all that is obsolete.

Long live the progressive Michurinist biological science!

Glory to the great Stalin, leader of the people and coryphaeus of progressive science!

Adopted unanimously at the Session of the V. I. Lenin All-Union Academy of Agricultural Sciences.

3

V. I. LENIN ALL-UNION ACADEMY OF AGRICULTURAL SCIENCES

Resolution on the Report of Academician T. D. Lysenko on the Situation in Biological Science (Read by P. N. Yakovlev)

Having heard and discussed the report of the President of the V. I. Lenin All-Union Academy of Agricultural Sciences, Academician T. D. Lysenko, "On the Situation in Biological Science," the Session of the Academy indorses completely the report in which an accurate analysis of the modern situation in biological science was made.

Two diametrically opposite movements have become defined

in biological science: one movement progressive, materialistic, *Michurinist,* named after its founder, the eminent Soviet naturalist and great transformer of nature, I. V. Michurin; the other movement reactionary-idealistic, *Weismannist* (Mendelian-Morganist), whose founders were the reactionary biologists, Weismann, Mendel, and Morgan.

The Michurinist movement proceeds from the point that new characteristics of plants and animals, acquired by them under the influence of the conditions of life, can be transmitted through heredity. The Michurinist doctrine equips practitioners with scientifically grounded methods for the systematic alteration of the nature of plants and animals, the improvement of existing, and development of new, varieties of agricultural crops and breeds of animals.

The Michurinist movement in biology is a creative development of Darwinist doctrine, and a new and higher stage of materialistic biology. Soviet agrobiological science, resting in its investigations on the notable doctrine of I. V. Michurin on plant development, and of V. R. Williams on soil development and procedures for the maintenance of conditions for a high productivity, and having found further continuation in the investigations of T. D. Lysenko and the entire collective of progressive Soviet biologists, has become a powerful weapon in the active, systematic transformation of animate nature. The Michurinist movement in biology renders daily aid toward the practice of socialistic agriculture. It is developing a new progressive agrobiological science, always increasingly expanding its aid to the collective and state farms, fighting for a greater productivity of socialistic agricultural production. The unity of theory and practice, as a prerequisite for successful understanding of the rules of development in animate nature, found full and clear embodiment in Michurinist agrobiological science. Thanks to this unity, modern agrobiological science has already made important advances in scientific knowledge and control over animate nature. There is no doubt that the future development of I. V. Michurin's doctrine will progressively augment the advances in subordination of nature to human will. The overwhelming majority of scientific workers in agricultural sciences follows the Michurinist path. Every kind of aid and support must be given these workers.

The Mendelist-Morganist movement in biology continues the idealistic and metaphysical doctrine of Weismann on the independence of an organism's nature from the external environment and on the so-called immortal "substance of heredity." The Mendelist-Morganist movement is apart from life and is practically fruitless in its investigations.

The Session of the V. I. Lenin All-Union Academy of Agricultural Sciences considers that the Michurinist movement, led by Academician T. D. Lysenko, has performed great and fruitful work in the exposure and destruction of the theoretical position of Mendelism-Morganism.

This work is of tremendous positive importance in the development of advanced biological science and agricultural practice.

The Session notes that until now the scientific-research work in a number of biological institutes and the instruction in genetics, selection, seed-growing, general biology, and Darwinism in the universities has been based on programs and plans permeated with ideas of Mendelism-Morganism, through which material injury has been done in the matter of ideological training of our cadres. In connection with this the general assembly considers necessary a radical reorganization of the scientific research work in the field of biology and a revision of the programs of educational institutions in the divisions of biological sciences.

This reorganization must contribute in equipping scientific workers and students with Michurinist doctrine. This is a requisite stipulation for the progress of the work of production specialists and in the investigation of actual problems in biological science. Simultaneously with a revision of the programs there must be organized a procedure for the creation of new, high-grade textbooks and issue of books and pamphlets devoted to the popularization of Michurinist doctrine.

The V. I. Lenin All-Union Academy of Agricultural Sciences must become the authentic scientific center of manifold and wide exploitation of Michurinist doctrine.

The Session of the Academy considers it necessary to subordinate investigations conducted in the institutes of the Academy to the problems of aiding collective farms, machine-tractor stations, and state farms carrying on the struggle for further elevation of the productivity of agricultural crops and animal-breeding.

The Session of the Academy calls upon the collective of scientific workers in agricultural sciences, all agronomists, zoötechnicians, and the leading people of the collective farm to unite more closely around the V. I. Lenin All-Union Academy of Agricultural Sciences, and in a united front, under the leadership of the party of Lenin-Stalin and of the great leader of the workers, teacher and friend of Soviet scientists, Iosif Vissarionovich Stalin, to develop the Michurinist doctrine, the advanced agrobiological science capable of the successful solution of the problems set before agricultural workers by our Party and Government.

X

Recantations

If ever an event justified the most vivid and expressive terms of which our language is capable, it is the event recorded in this chapter. Nothing remotely resembling it has happened in over three hundred years. In fact, no precedent at all can be found for such an abjuration of a well-established and tested science in favor of an archaic fallacy. When we consider the practical importance of the knowledge rejected and the bigotry and stupidity shown by the majority of the rapidly promoted but uneducated academicians, the whole affair becomes preposterous. Victories of majority opinion are common but, when accompanied by the forced humiliation of enlightened and superior persons, they have a habit of turning sour. Groups which win such victories sooner or later find themselves acutely embarrassed and find the time not distant when they would like to forget the whole thing. Explaining away such triumphs becomes one of the more trying tasks of apologetics.

Before we condemn the Russian geneticists who recanted, we should endeavor to learn all the circumstances which conditioned their acts. We should try to put ourselves in their places and ask ourselves what we should do if we and our families were among the physical possessions of the Communists. We should ask also how we would behave if, for the preceding thirty years, we had had access only to such carefully culled and slanted sources of information that our intellects had become warped and our consciences had, in consequence, become the property of the Party? We might know, as we recanted, that we were deserting the truth but this might seem to us a minor peccadillo compared with our not conforming to the all-righteous Party line.

We do not, as a rule, admire those who recant but we must remember that conditions have arisen in which individuals who rank among the greatest of our species have renounced what they considered the truth. Joan of Arc recanted temporarily but recovered her courage and died at the stake, while Galileo recanted permanently and thus was able to die a natural death.

Stubborn folk who do not recant often achieve martyrdom, and we scientists have before us the examples of Michael Servetus and Giordano Bruno. But science cannot as yet compete with religion in either the number or the quality of its martyrs. Religion has frequently produced men who were quite willing to trade the passing discomfort of a messy end on earth for a glorious immortality elsewhere, but scientists tend to weigh the objective evidence for future bliss along with the immediate consequences of their alternate courses of action. As a result perhaps their decisions tend to be more practical than heroic. There do not seem to have been any especially desirable mansions in Valhalla reserved for those who lose their lives defending Saracenic learning, Copernican astronomy, or Mendelian genetics.

Five Russian geneticists disavowed their science because the Michurinists seemed to have their hearts set on a unanimity of biological opinion and also the implements for achieving their desires. The motives of the five were certainly not identical. Zhdanov's confession of error is probably inspired purely by political considerations. His letter to Stalin is good natured, shows no mental stress or strain, and even contains a few bits of subtle irony which he knew the authorities would never spot. On the other hand Zhukovski, Alikanyan, and Polyakov, who admitted their sins publicly during the tenth session of the Lenin Academy, probably were carried away, at least in part, by the general mob hysteria although they were certainly aware that their action promised them personal safety. Zhebrak, who wrote out his recantation two days after the meetings ended, simply put his loyalty to Communism before his scientific integrity. Petrov (Ch. VI, § 1) had joined the Michurinists earlier when he spoke during the third session on August 2. He seems to have been received into the orthodox sodality.

We have no records yet of any recantations of Rapoport, Zavadovski, Schmalhausen, or Nemchinov.

1

GEORGE ZHDANOV

Letter to the Central Committee of the Communist Party of the Soviet Union (Bolshevik): To Comrade Stalin

(From *Pravda,* August 7, 1948)

[*The following letter contains George Zhdanov's confession of his most serious error in supporting Mendelian genetics. It was dated July 10, 1948, but was not published until August 7, the day the Session of the Lenin Academy came to an end. How beautifully its publication was timed is shown by its obvious effect upon Zhukovski and Alikhanyan who recanted on the very day it appeared. George Zhdanov was the son of Col. Gen. Andrei A. Zhdanov, a member of the Central Committee, whose sudden death was announced on September 1. According to Nicolaevsky,[1] the first symptom of General Zhdanov's eclipse by Malenkov was the fact that his son was forced to make this recantation. It is interesting to note that this confession of error in a technical scientific field is addressed to Stalin himself.*]

Appearing at a seminar of lecturers with a paper on the debatable questions of modern Darwinism, I undoubtedly committed a whole series of grave mistakes.

1. The very presentation of this paper was a mistake. I clearly underestimated my new position as a worker within the apparatus of the Central Committee, underestimated my responsibility, and failed to take into consideration that my appearance will be thought of as an official point of view of the Central Committee. Here was expressed that "university habit," when I made known my viewpoints in any scientific controversy without meditation. Therefore, when I was invited to give a paper at a seminar of lecturers, I decided to express my deliberations there with the reservation that this is "a personal point of view," so that my appearance would not obligate anyone to anything. There is no doubt that this was "professorial" in its poor judgment and not the Party position.

2. The fundamental error in the report itself was its bearing on a reconciliation of conflicting ideas in biology.

Representatives of formal genetics began to appear from the very first day of my work in the department of science with complaints that new varieties of useful plants (buckwheat, kok-saghyz, geranium, hemp, citrus fruit) produced by them and possessing improved qualities are not being introduced into production and are meeting opposition from the followers of Academician Lysenko. These undoubtedly useful forms of plants were produced by means of the immediate influence of chemical and physical factors on the germ cell (the seed). The Michurin doctrine does not deny the possibility and expediency of such an influence, acknowledging the presence of many other more important ways of altering an organism. Formal geneticists consider their own methods (strong stimulus of an organism with X-rays, ultraviolet rays, colchicine, acenaphthene) the only ones possible. I myself account for it that the mechanism of influence in these

[1] Boris I. Nicolaevsky, "Palace Revolution in the Kremlin," *New Leader* 32:(12):8. March 19, 1949.

agents can and must be explained not by formal but by Michurin genetics.

My error lay in that, having decided to defend these *practical* results which appeared "as Danaean gifts," I failed to subject the fundamental methodological defects of Mendel-Morgan genetics to a merciless criticism. I realize that it is a utilitarian approach toward practical work, a "pursuit after a kopeck."

The conflict of ideas in biology frequently assumes abnormal forms of squabbles and scandal. However, it seemed to me that in general there was nothing here apart from these squabbles and scandals. Consequently, I failed to appraise the principal aspect of the problem, approached the controversy nonhistorically, and failed to analyze its deep roots and causes.

All this taken collectively produced a tendency "to reconcile" the disputing sides, to erase discords, to emphasize that which unites and not that which separates the opponents. But in science, as in politics, principles do not reconcile themselves but conquer;[2] the conflict does not proceed along a path of concealing, but of exposing contradictions. The attempt to reconcile principles on a basis of practicality and narrow pragmatism, and inappreciation of the theoretical aspect of the controversy led to eclectics, to which I confess.

3. My sharp and public criticism of Academician Lysenko was an error. Academician Lysenko is now the recognized leader of the Michurin influence in biology; he defended Michurin and his doctrine from the attacks of bourgeois geneticists; he personally contributed much to science and practice in our economy. Taking all this into consideration, a criticism of Lysenko, of his individual deficiencies, ought to be conducted so that it would not weaken but strengthen the positions of the Michurinists.

I do not agree with some of the theoretical attitudes of Academician Lysenko (denial of intraspecies struggle and mutual aid, underestimation of the internal specificity of an organism); I think that as yet he has made poor use of the treasure depository of Michurin teachings (to wit, that is why Lysenko has not brought forth any important varieties of agricultural plants);[3] I think that his guidance of our agricultural science is weak. The V. I. Lenin All-Union Academy of Agricultural Science, which he heads, is working far from its full power. No works at all are being performed in it on animal-breeding or on the economics and organization of agriculture, and the work in agricultural chemistry is poor. However, all these deficiencies should not have

[2] This gives us a Russian view of both science and politics. It probably tells us more than was intended.

[3] Again the truth slipped out.

been criticized as I did in my report. As a result of my criticism of Lysenko the formal geneticists became "third rejoicers."

I, who am dedicated with all my heart and soul to Michurin's teaching, criticized Lysenko not because he is a Michurinist but because he is developing Michurinist principles insufficiently. Thus, objectively, the Michurinists lost while the Mendelians-Morganists won from such criticism.

4. Lenin frequently said that a recognition of the necessity of any phenomenon conceals within itself the risk of falling into objectivism. To a definite extent, I likewise did not escape this danger.

I characterized the place of Weismannism and Mendelism-Morganism (I do not distinguish between them) to a considerable extent according to Pimen:[4] both good and evil noticing with indifference. Instead of a sharp turn against these antiscientific views (expressed among us by Schmalhausen and his school), which in theory appears as a behind-the-veil form of popishness, of theological representations of the origin of species in consequence of separate acts of creation, but in practice lead toward "finality" and to a denial of the ability of man to alter the nature of animals and plants, by mistake I set before myself the task "to realize" their place in the development of biological theory and to find "the rational seed" in them. As a result my criticism of Weismannism was weak and objectivist, and essentially—shallow.

As a result, again the main blow came to Academician Lysenko, that is by ricochet on the Michurin course.

Such are my mistakes as I understand them.

I consider it my duty to assure you, Comrade Stalin, and through your person the Central Committee of the Communist Party of the Soviet Union (Bolsheviks), that I have been and remain an ardent Michurinist. My mistakes result from my insufficient discrimination of the history of the problem and that I built an incorrect front of struggle for the Michurin doctrine. All of this was due to inexperience and immaturity. I will correct my mistakes with deeds.

July 10, 1948 GEORGE ZHDANOV.

[4] This is a reference to a well-known character in Russian literature. Pimen (a monk) appears in Pushkin's *Boris Godunov*.

2

P. M. ZHUKOVSKI'S RECANTATION

(From *Proceedings* of the Lenin Academy, 1949)

Academician P. P. Lobanov: Academician P. M. Zhukovski wishes to make a statement.

Academician P. M. Zhukovski: Comrades, late last evening I decided to make the following statement. I mention late last evening purposely because I did not know that Comrade G. Zhdanov's letter will appear in today's *Pravda* and, therefore, there is no connection between my statement and Comrade G. Zhdanov's letter. I think that the representative of the Ministry of Agriculture Lobanov can confirm this, as I asked him over the telephone last evening to allow me to make a statement at the session today.

In man's life, particularly in our historic days, there occur moments of great moral, philosophical, and political significance. Such moments I experienced yesterday and today. My appearance of two days ago was unsuccessful and was my last appearance, as it is said here, against Michurin, although I never before appeared against Michurin's teaching personally. In addition, it was a final appearance from an incorrect biological and ideological position. *(Applause)*

The unfortunate history of my article "Darwinism in an Oblique Mirror," and our President's reply to this article, took me subsequently from the field of ideological struggle to that of personal offense. True, I remain as before with the concept of the existence of intraspecific competition. But I must say that it was precisely in this period that my relations with the President became considerably strained.

My appearance of two days ago, when the Party's Central Committee marked the dividing ridge which separates the two movements in biological science, was unworthy of a Soviet scientist and member of the Communistic Party.

I admit that I took an erroneous stand. Yesterday's remarkable speech of Academician Lobanov, his phrase addressed directly to me. "We are not on the same path with you," and I regard P. P. Lobanov as a great statesman—these words upset me greatly. His speech threw me into confusion. A sleepless night helped me think over my behavior.

Academician Vasilenko's appearance likewise produced a great impression on me, because he showed how closely Michurinists are linked with the people and how important it is in this period to defend the President's authority.

The exceptional unity of the members and guests at this session, the demonstration of the strength of this unity and closeness with the people and, on the other hand, the demonstration of the opposition's weakness are so apparent to me that I declare: I will fight—and sometimes I am very capable at it—for the Michurinist biological science. *(Prolonged applause)*

I am a responsible individual because I am in the Committee for the Stalin Awards in the Council of Ministers and in the Expert Commission for the awarding of higher scientific degrees. Thus, I think that I have a moral duty—to be an honest Michurinist, to be an upright Soviet biologist.

Comrade Michurinists! If I have declared that I am entering the ranks of the Michurinists and that I will defend them, then I do it honorably. I appeal to all Michurinists, among whom are both my friends and my foes, and declare that I will honestly fulfill all that I have stated here today. *(Applause)*

I am sure that, knowing me, it will be also believed in the present instance that I did not make my declaration out of cowardice. An important trait of my character in life has always been a great impressibility. All know that I take everything very nervously. Therefore you will believe me that the present session has truly made a tremendous impression upon me.

It is said here (and it is a justified reproach) that we do not carry on a struggle in print against foreign reactionaries in the field of biological science. I declare here that I will wage this war and will give it political significance. I think that it is necessary, finally, that the voice of Soviet biologists be heard on the pages of our scientific publications on the great ideological abyss that separates us. And only the alien scientist who will understand that the bridge must be crossed toward us, and not toward him, may depend on our attention.

Let the past which separated us from T. D. Lysenko (true, not always) pass into oblivion. Believe me, that today I make a Party step and appear as a true member of the Party, i.e., honorably. *(Applause)*

In addition, I declare that Academician Vasilenko's appeal to protect the President's prestige will be fulfilled by me. *(Applause)*

3

S. I. ALIKHANYAN'S RECANTATION

(From *Proceedings* of the Lenin Academy, 1949)

Academician P. P. Lobanov: Comrade S. I. Alikhanyan wishes to make a statement.

S. I. Alikhanyan: Comrades, I requested permission of the chairman to speak, not because I read George Andreevich Zhdanov's declaration in today's *Pravda.* I resolved to make a declaration last night, and the representative of the Ministry of Agriculture P. P. Lobanov can verify that I had a conversation with him about this yesterday, August 6.

I have followed this session very attentively and have experienced much during these days. It follows from all that has taken place here at this session that I, a young Soviet scientist who has thought all this over, must as a scientist draw a fundamental conclusion. The matter proceeds, comrades, and I appeal to my adherents—

N. G. Belenki: Past or present?

S. I. Alikhanyan: Both past and present. The matter concerns the struggle of two worlds, a struggle of two world outlooks, and there is no need for us to cling to old conditions which were presented to us by our teachers.

We yielded greatly to polemic passions, which were kindled in this discussion by our teachers. Because of the controversy we were unable to see the new, rising movement in the science of genetics. This new movement is Michurin's teaching. And, as I have already mentioned, it is important for us to understand that we must be on the same side of the scientific barricades[5] as our Party and our Soviet science.

It would be naïve to think that we are required to refuse everything positive and useful accumulated in the entire course of science's development. A refusal of all that is reactionary, untrue, and useless is required of us. And we must accomplish this sincerely and honestly as befits modern scientists.

I call upon my comrades to draw very serious conclusions from what I say. I, as a Communist, cannot and must not obstinately contrast, in the ardor of controversy, my personal views and conceptions against the entire progressive course in the development of biological science.

[5] This is called particularly to the attention of American apologists.

On leaving this session the first thing that I must do, is to revise not only my relationship to the new, Michurinist science, but likewise all my former scientific activity. I call upon my comrades to do the same.

I cannot think of my existence without active and useful activity for the good of Soviet society and Soviet science. I believed in our party and our ideology when I went into battle with my soldiers. And today I sincerely believe that, as a scientist, I act honestly and truthfully and go with the Party and with my Country, and if you, comrades, do not do likewise, you will be found lagging behind, you will fall behind in the progressive development of science. Science does not tolerate indecision[6] and lack of principle.

From tomorrow, I will begin to rid not only my own scientific activity of old, reactionary, Weismannist-Morganist views, but will also begin to remake, to break in two, all of my students and comrades.[7]

It is impossible to conceal that this will be an unusually difficult and painful process. It is possible that many will not understand it; well then, it cannot be helped, they are not with us. It means they will not be able to appraise correctly the help that was shown us by the Party at the fundamental turning point which took place in science, and will not be able to understand that the matter is not a question of mere differences in individual principles.

I will strive so that my comrades will not expend idly all their experience and knowledge or leave them in laboratories, but will spread them out widely into the national economy. This should not be difficult if, freed from the weight of unnecessary metaphysical concepts, we proceed honestly toward this end in close friendship with all the scientists of our country.

And only in our country, a country with the most advanced progressive world outlook, can the seedlings of the new scientific movement develop, and our place is with the new and advanced. And I, for my part, categorically declare to my comrades that henceforth I will fight my adherents of yesterday who will not understand and join the Michurinist movement. I will not only criticize the vicious Weismannism-Morganism which appeared in my works, but will also take an active part in the progressive movement forward of Michurinist science.

I am convinced that the collective of biologists of the Moscow State University will understand me correctly and that we will turn the foremost university of our country—the Moscow

[6] This is in startling contrast to Western scientific standards.

[7] A method of securing his personal safety?

State University—into a center of propaganda of Michurin's teaching, a center for the exploitation of Michurinist biology. *(Applause)*

4

I. M. POLYAKOV'S RECANTATION

(From *Proceedings* of the Lenin Academy, 1949)

Academician P. P. Lobanov: Professor I. M. Polyakov wishes to make a statement.

I. M. Polyakov: Comrades! Last evening in conversation with my friends, present at today's session, I said that this session was a very great and stirring event in my life, compelling considerable revaluation.

You comrades know of my numerous appearances at scientific assemblies and conferences and my scientific articles and textbooks. I have always tried to make an honest analysis of the many outstanding and important problems in the theory of evolution, Darwinism, and genetics. I have endeavored to reason out the fundamental theoretical thesis in our science from Marxist-Leninist positions and to criticize sharply the reactionary views of foreign and some of our scientists. You comrades know that I have acted in this way for many years.

But when I was reproached from this platform that my appearance here had been slipshod, I was reproached properly. Not to proceed to the end but to take an intermediate attitude is an act unworthy of a Bolshevik-scientist. One must form one's position clearly and precisely. One must state directly that the Michurinist movement is the general course of development for our biological science and that it is necessary to follow this course. It is the only course possible for Party and non-Party Bolsheviks who want to work in the sphere of our biological science, bringing benefit to our Soviet people and our Fatherland.

I wish to note that during the past eight to nine years I and my closest collaborators have worked on the problem of selective fertilization—one of the most important problems of Michurinist genetics. We arrived at a number of interesting conclusions; these conclusions completely confirm the Michurinist point of view. I wrote about this in works which are now being published.

But it is impossible to stop on that; it is necessary to make further deductions. One must be logical and not attempt to connect things that are disconnected.

It is a most serious matter for a scientist who is seriously occupied in his work and who likes his work to be reconstructed along a "special course." I still must think long and seriously over many problems. We can and should argue many questions in our science fruitfully and creatively. If, for instance, we argue about the struggle for existence and selection there is nothing wrong in this, since comradely arguments among Soviet scientists on this or that concrete question of science can only be useful. But it is necessary to understand what is primary and fundamental—that our Party helped us to produce a deep and radical break in the realm of our science and showed us that Michurinist teaching determines the basic line of development of Soviet biological science and that it is necessary to make a conclusion and to work from here, developing the Michurinist movement.[8] And it must be demonstrated by one's work and not by mere declarations. This must be the program of my work as a Communist-scientist. And if this course is not taken, then whether one wants it or not, he will attract to himself people inclined to unprincipled nonsense, people who do not see anything fundamental being done in our country beyond the private scientific controversies. I call upon all our Soviet biologists to reach determinedly toward the same conclusions which I reached.[9] It will not be easy or simple for many to do and it is necessary to think everything over very deeply, but, I repeat, one must make an absolute rupture with false views and resolutely criticize the metaphysical, idealistic, Weismannist views of foreign reactionaries from science and the echoes of these views in the works of certain Soviet scientists. We must help our Party to expose this reactionary, pseudoscientific rot which is spread abroad by our enemies. We must understand that this rot has influenced certain Soviet scientists and that it must be completely uprooted. The Michurinist movement in science, led by T. D. Lysenko, is a broad and great national scientific movement, a movement which promotes a more rapid passage on the great path of the triumphant construction of communistic society. The workers of Soviet biological and agricultural science must work within this movement. I, likewise, will work in this movement, applying all my effort for the development of the great Michurinist teaching. [No applause was reported for Polyakov.]

[8] A confession of political dictation to science.

[9] Perhaps they might succeed through an exercise of Russian objectivity.

5

A. R. ZHEBRAK'S RECANTATION

(From *Pravda,* August 15, 1948)
To the Editorial Office of Pravda

Please publish the following text of my declaration:

As long as both courses in Soviet genetics were recognized by our Party and the controversies concerning these courses were considered as creative discussions of the theoretical problems of modern science, furthering the search for truth through controversy, I persistently defended my views which differed on particular questions with the views of Academician Lysenko. Now, however, since it has become clear to me that the fundamental aspects of Michurin's direction in Soviet genetics are approved by the Central Committee of the All-Union Communist Party (Bolsheviks), then I, as a member of the Party, do not consider it possible for me to retain those views which are recognized as erroneous by the Central Committee of our Party.

I consider the evaluation of Weismannism by Academician Lysenko as an idealistic movement in biology absolutely correct. His criticism of the Weismann concept on somatic and germ plasm agrees with my criticism, which I made in different publications (1936-37) as a result of my experiments on the alteration of crossing-over. (See *Socialistic Reconstruction of Agriculture,* Nos. 8 and 12, 1936, and *Reports of the Academy of Science, USSR,* biological series, No. 1, 1937.) I cite several references to my old articles: ". . . the very arrangement of the experiments (Weismann's) was so primitive that it was impossible to base on them the theoretical deductions which were ascribed to them. . . ."

". . . the Weismann concept was dialectic from the viewpoint of dialectical materialism because it divided the organism into two substances and excluded a mutual exchange between the substances of one and the same organism. Yet more amazing is the fact that this concept has survived so long in genetics and that even now a certain segment of geneticists adheres basically to the same methodological attitudes as were held by Weismann's followers several decades ago. . . ." Furthermore, I demonstrated the falseness of Weismann's theory and of a group of other theories logically resultant from it, particularly the theory of phenotype and genotype.

My views likewise do not conflict with Michurin's concept on the question of the influence of external environment on the organism. In 1936 my views were stated as follows:

> We regard the phenotypic changes (acquired characteristics) as hereditary, whether they are produced by developmental modifications or by change in environment. They are hereditary in the sense that the germ cells originate in a concrete phenotype, in a concrete modification, and once the phenotype of the organism is delineated, then also the processes which are taking place become distinguishable, and consequently, these variations must be reflected in some manner in the germ cells which subsequently may give origin to altered organisms.

The viewpoint expressed in my work is opposite in its methodological relationship to that of Weismann and Morgan. My conception was based on a large number of experiments. I might add that my work was introduced for publication by Academician V. R. Williams.

My work in obtaining new amphidiploidic forms of wheat by a method of remote hybridization and action of chemical elements involves recognition of the dependency of the hereditary foundation to the external environment. This idea was expressed in a work in 1944 in this manner:

> It is of particular importance in all our works that the results obtained in the creation of new amphidiploidic types of wheat were obtained by means of the reaction on the hereditary foundation of the plant of such an external factor as colchicine. This completely destroys the views of the old autogeneticists who considered the germ plasm as isolated from external conditions and resistant to external influences. (*Synthesis of New Wheat Types*, page 43. State Publishing House for Agriculture, 1944.)

I have brought forward these references in order to show that I have not been concerned with either the idealism of Weismann or with that of the autogeneticists. In the discussion of 1936 I came forward with my own distinct point of view, which developed differences with both one and the other side of the controversy.

In my scientific work I have endeavored and am endeavoring to realize Michurin's motto: "Do not expect favors from Nature; to take them from her—is our task," and the motto of Marx: "Not only to explain the world but also to remake it." Man indisputably can create better plant forms than nature.

The actual achievements of our laboratory, which constitute the priority of our country, serve as proof that I made use of effective methods in the alteration of plants and the development of polyploidic types of wheat and other cultivated plants. In my scientific work I use Michurin's method of remote crossing-over, a method of interbreeding geographical races, a method of the reaction of chemical and physical factors, and so on.

Since it is the sacred duty of the scientists of our Country to keep in step with the entire nation, to fulfill the vital needs of our Government and the demands of the people, to fight the vestiges of capitalism, to aid the communistic education of the workers and to advance science continuously forward, then I, as a member of the Party and a scientist of the people, do not want to be considered a renegade, do not want to be aside from the fulfillment of the noble tasks of the scientists of my Fatherland. I want to work within the framework of that structure which is acknowledged foremost in our country and with those methods which were propounded by Timiryazev and Michurin. Henceforth, I shall endeavor with all my strength toward the aim that my works will be of the greatest benefit to our Country, that they will creatively develop the legacy of Timiryazev and Michurin, and that they will aid in the communistic construction of our Fatherland.

I think that my experimental works in the alteration of cultivated plants, generalized on the foundation of Timiryazev's and Michurin's theory, will add their mite to the development of Soviet biological science.

PROFESSOR A. R. ZHEBRAK.

August 9, 1948

6

EDITORIAL COMMENTS ON ZHEBRAK

(*Pravda,* August 15, 1948)

In publishing the letter of Professor A. R. Zhebrak, the editorial office of *Pravda* cannot agree with a series of his inaccurate statements.

1. His statement that "both directions in Soviet genetics were recognized by our Party," that is, the progressive materialistic Michurin teaching and the reactionary-idealistic Weismann concept in biology, is incorrect.[10] Dialectic materialism as the foun-

[10] This statement is false. The record shows that, for a while, the party did support both the Lysenko and the Vavilov groups (page 41). Indeed, some reputable genetics research was supported until 1946 (Dubinin, *Science* 105:109-12. 1947).

dation of our world outlook has always led an irreconcilable, determined struggle against all varieties of idealism in the field of social as well as of the natural sciences. The Party therefore supports the Michurin teaching, which is based on the secure positions of dialectic materialism, truthfully reflecting objective truth.

2. Incredible and pretentious is the statement of Professor A. R. Zhebrak that the criticism of Weismannism in Academician Lysenko's account "On the Situation in Biological Science" coincides with the criticism which Zhebrak made in his works, indicated in the letter. In reality, A. R. Zhebrak in his works on the fundamental problems of biology supports Weismann-Mendel-Morgan. In these works A. R. Zhebrak eulogizes and exalts Mendelism-Morganism. "The Morgan school, he writes, enriched genetics with new ideas and raised it to a higher theoretical level." ("Some Contemporary Problems of Genetics," *Socialistic Reconstruction of Agriculture,* No. 8, 1936, p. 103.)

In works indicated by himself, the author of the letter supports Weismann attitudes on the fundamental problems of biology. He states that "the development of biology and particularly of that part which occupied itself with the study of heredity and variability—of genetics—accumulated a great mass of experimental data against the inheritance of acquired characteristics. . . . Both germ cells and the historic influences, which were tested by organisms in the process of their individual life, were involved in the historical formation of species. It is quite clear that the guiding and fundamental role in this process belonged to germinal elements." (*Ibid.,* page 99.)

It is clear that in the light of similar statements A. R. Zhebrak's declaration that he was not implicated in Weismannism-Morganism, is inaccurate and false.

3. In his letter Professor Zhebrak endeavors to prove the undemonstrable, namely, as though he made use of Michurin's methods and mottoes in his own works. However, it is sufficient to quote but one of Professor Zhebrak's characteristic statements with a disdainful evaluation of the works of I. V. Michurin in order to expose the entire falseness of his assertions. Thus, for example, in the article "Categories of Genetics in the Light of Dialectic Materialism," Professor Zhebrak wrote: "Regardless of the great merits of some of our outstanding selectionists, among them of Ivan Vladimirovitch Michurin, we cannot entirely cloak them with authority." (*Ibid.,* No. 12, 1936, p. 86.)

At the same time, as it has been shown, A. R. Zhebrak exalted the founders of Mendelism-Morganism in every way. Thus, it appears that A. R. Zhebrak, stating that he cannot support those positions which have been condemned by our Party, simultaneously endeavors to show that he did not support them. Such inconsistency does not suit an individual desiring to condemn his errors honorably and openly.

4. Professor A. R. Zhebrak shows a tendency to contrast the teaching of I. V. Michurin with the works of Academician T. D. Lysenko and other Michurinists. Comrade Zhebrak ignores the fundamental problems of the struggle with Mendelism-Morganism, first of all—the problem of the inheritance of properties and characteristics by plants and animals, acquired under the influence of environment. He endeavors to demonstrate as though the Mendel-Morgan views do not disagree with Michurin's teaching. It is necessary to recognize these maneuvers of Comrade Zhebrak as an attempt with worthless means.

5. With regard to the boastful statements of Professor Zhebrak on his supposed "real achievements" in the field of the alteration of plants, it is now generally known that in reality no tangible achievements have been made by Zhebrak and his laboratory. As yet not a single valuable variety has been developed by Zhebrak; not a single, practically valuable suggestion for our agriculture has been offered. As with the other representatives of the Weismann-Morgan orientation, Professor Zhebrak is like a sterile fig tree.

6. The Editor takes into consideration the declaration of A. R. Zhebrak that he does not desire to be considered a renegade, and promises to aid in the construction of Communism by creative developing of the heritage of Timiryazev and Michurin. The subsequent work of Professor A. R. Zhebrak will demonstrate to what extent this declaration is sincere.

XI

The Heresy Hunt Spreads

The decision of the Central Committee of the Communist Party to approve Lysenko and all that he stands for, together with the Resolution adopted by the All-Union Lenin Academy of Agriculture, effectively wiped out genetical research in Russia. Indeed, ever since 1940 when Vavilov was arrested, genetics had been under the cloud of official displeasure, but it had been allowed to linger on in certain institutions, though clearly it had a most dubious future. Students in the universities were advised not to become geneticists and, as the older scientists died, there were few to replace them. As late as 1947, Dubinin (see Bibliography) described some real genetic work still in progress, but in September of that year both he and Zhebrak were attacked most vigorously by Laptev (Ch. III, § 7), who obviously acted with official approval. The Lysenko clique had at last reached the point where it could stage one final offensive and mop up the remnants of that great group of scientists who had been assembled by Vavilov.

During all these years, however, the Michurin quackery made little progress in the Russian universities. The biological departments were immune to the infection, and we can easily understand why. Students intelligent enough to pass courses in physiology, morphology, or ecology were able to understand the basic principles of genetics. Being in contact with the basic standards of scientific research, they could hardly have been impressed by such academicians as those whose speeches we have recorded in the earlier chapters. A stray Michurinist wandering over from some department of dialectical materialism into a group of biologists would definitely feel out of place and unwelcome. The fact should also be recorded that the universities' administrative authorities were too intelligent and scholarly to be fooled by the charlatans. But all this was now to be changed.

When Weismannists-Mendelists-Morganists were declared by official pronouncement to be enemies of Communism, the presidents and deans of the universities and the commissars of education suddenly found themselves to be gravely "in error" in that they had not been sufficiently vigilant and had failed to expose mercilessly the representatives of bourgeois science. Apparently even the august Academy of Science itself had been tolerating the intolerable. The only course of action obviously was for these negligent officials to confess dereliction, promise to reform, and hope for forgiveness.

Fortunately, a standard technique has been devised for those who err and repent. For important groups or individuals it consists of an extravagantly eulogistic letter to Stalin with a few personal grovels thrown in as a conventional token of penance. Two such letters are included in this chapter, a letter from the Praesidium of the Academy of Science (§ 1) and a letter from the Praesidium of the Academy of Medicine (§ 5). The Praesidium of the Academy of Science also passed a resolution giving the details of a proposed purge (§ 2). An editorial in the same issue of *Pravda* (August 27, 1948) from which sections 1 and 2 are taken constitutes section 3. It shows what scientists may expect in a police state. S. Kaftanov, Minister of Higher Education, wrote an editorial in *Izvestia* (September 8, 1948), and here he shows how he will make up for lost time (§ 4).

1

PRAESIDIUM OF THE ACADEMY OF SCIENCES, USSR

Letter to Stalin

To Comrade I. V. Stalin
(Four-column spread in *Pravda,* August 27)
Dear Iosif Vissarionovich:

The general meeting of the Praesidium of the Academy of Sciences, USSR, dedicated to the consideration of the question concerning the status and problems of biological science in the Academy of Sciences, USSR, greets you, our beloved leader and teacher, with cordial Bolshevist greetings and gratitude for the interest and aid which you daily show Soviet science and Soviet scientists. Soviet science is indebted to you for its great achievements; you have always directed the development of science in the interests of the people. You have always helped and are helping us to overcome reactionary teachings, hostile to the people, and to protect advanced Soviet science from the danger of its estrangement from the people and from practical work.

The recently concluded session of the V. I. Lenin All-Union Academy of Agricultural Sciences, and the report of Academician T. D. Lysenko "On the Situation in Biological Science," approved by the Central Committee of the Communist Party of the Soviet Union (Bolsheviks), are of decisive importance for the future development of biological science.

We have before us a problem of exceptional importance—to develop Michurinist biological science and to exterminate the anti-national, idealistic, Weismannist-Morganist movement in biology.

The great representatives of natural science in our country: Sechenov, Pavlov, Mechnikov, Timiryazev, Michurin, Dokuchayev, and Williams were fighters for a progressive biological science, and defended the materialistic theory of development of the organic world. Our great compatriot, Academician I. V. Michurin, created a new epoch in the development of materialistic biology. The official science of old Russia was incapable of appreciating the great transformer Michurin and his teaching. Michurin was discovered for our people and for progressive science by the genius of Lenin and Stalin.[1] In a Soviet country, under conditions of triumph of the socialistic order, Michurin's teaching has brought remarkable results and has gained great possibilities for its development.

The reactionary course in biology-Weismannism-Morganism was formed in opposition to revolutionary Michurinist teaching.[2] This idealistic movement, prevalent in capitalistic countries, unfortunately found supporters among a number of Soviet biologists, and especially among certain scientists in the Division of Biological Sciences of the Academy of Sciences, USSR.

The Praesidium of the Academy of Sciences, USSR, and the Bureau of the Division of Biological Sciences made a grave error in giving support to the Mendelist-Morganist movement to the detriment of progressive Michurinist teaching. The Praesidium of the Academy of Sciences inadequately directed the biological institutes of the Academy of Sciences and retained opponents of Michurinist teaching in guiding positions. As a result, the scientific institutes of the Division of Biological Sciences contributed very little to the solution of the practical problems of socialistic construction.

[1] Lenin undoubtedly initiated Michurin into the Apostolic Succession.

[2] Weismann long antedated Michurin. When Morgan made his fundamental discoveries, Michurin, an obscure importer of plants, was unknown to the scientific world. To state that the theories of Morgan were formed in opposition to Michurin is nonsense.

The Praesidium of the Academy of Sciences promises you, dear Iosif Vissarionovich, and through you, our Party and Government, determinedly to rectify the errors we permitted, to reorganize the work of the Division of Biological Sciences and its institutes, and to develop biological science in a true materialistic Michurinist direction.

The Academy of Sciences will take all necessary measures in order to give Michurinist biological science full development in biological institutes, journals, and publishing activity. The biological institutes of the Academy of Sciences are revising the programs of their scientific investigations in order to bring closer the works of the institutes to the needs of the country's national economy. Work on the theoretical generalization of the achievements of Michurinist biology, and exposure of the reactionary "theory" of the Weismannists-Morganists, will occupy the official position in the plans of the biological institutes.

We promise you, Comrade Stalin, in the name of the great objectives of our people, in the name of Communism's victory, to take the leading position in the struggle against idealistic, reactionary teachings, and to clear all paths for an unhindered development of progressive Soviet science.

PRAESIDIUM OF THE ACADEMY OF SCIENCES, USSR

2

PRAESIDIUM OF THE ACADEMY OF SCIENCES, USSR

Resolution, August 26, 1948, on the Question of Status and Problems of Biological Science in the Institutes and Institutions of the Academy of Sciences, USSR

(From *Pravda,* August 27, 1948)

The Session of the V. I. Lenin All-Union Academy of Agricultural Sciences (LAAAS) has placed a number of important problems before Soviet biological science, whose solution must contribute to the great work of socialistic construction. The Session of the LAAAS has revealed the reactionary, antinational nature of the Weismann-Morgan-Mendel movement in biological science, and has exposed its actual bearers. The destruction of the anti-Michurinist movement has opened new possibilities for the creative development of all branches of advanced biological science.

The materials of the LAAAS Session have shown, with all transparency, that there has been in progress a struggle between two diametrically opposite, according to their ideological and theoretical concepts, movements in biological science: the struggle of a progressive, materialistic, Michurinist movement against a reactionary, idealistic, Weismannist-Morganist movement.

The Michurinist movement, having creatively enriched the theory of evolution and revealed the laws of development of living nature, has through its methods of controlled alteration of the nature of plants and animals made an outstanding contribution to the practice of socialistic agriculture. The Weismannist-Morganist movement, maintaining the independence of hereditary changes of an organism from its characteristics of form and its conditions of life, has supported the idealistic and metaphysical views, torn apart from life; has disarmed practical workers in agriculture from their goal of improving existing and creating new varieties of plants and animal breeds; and has occupied itself with fruitless experiments.

The Academy of Sciences not only failed to take part in the struggle against the reactionary bourgeois movement in biological science, but actually supported representatives of formal-genetic pseudoscience in the Institute of Cytology, Histology, and Embryology; in the Institute of Morphological Evolution; in the Institute of Plant Physiology; in the Main Botanical Gardens; and in other biological institutions of the Academy of Sciences.

The Praesidium of the Academy of Sciences, USSR, admits that its work in directing the Academy's biological institutes was unsatisfactory.

Under a cloak of objective relationship to the two opposing movements in biological science, support was actually given to the reactionary Weismannist trend in the operation of the Praesidium of the Academy of Sciences and the Division of Biology, while the only true trend, the Michurinist, was slighted and circumscribed in every way.

The Praesidium of the Academy of Sciences permitted a systematic strengthening of the supporters of the pseudoscientific, Weismannist, formal-genetic movement. The Praesidium of the Academy of Sciences, USSR, made an error, by having raised the question in 1946 of the establishment of a special Institute of Genetics and Cytology in opposition to the existing Institute of Genetics, directed by Academician Lysenko. Representatives of Morganist genetics concentrated themselves, with the Praesidium's support, in a number of the biological institutes of the Academy of Sciences.

The Bureau of the Division of Biological Sciences and its director, Academician L. A. Orbeli, were unable to place the theoretical work of the biologists of the Academy of Sciences at the service of the urgent problems of socialistic construction in the fields of plant cultivation and animal-breeding.

The struggle against the bourgeois, reactionary currents in biology was conducted entirely inadequately in the institutes of the Division of History and Philosophy. Thus, the Institute of Philosophy of the Academy of Sciences, USSR, required to conduct a systematic struggle against all manifestations of idealism in science and to defend the materialistic world outlook, failed to give the necessary support to the Michurinist, materialistic movement in biology and did not occupy itself with the scientific elaboration of the methodological problems of biological science.

The Editorial-Publishing Council and the editorial boards of the journals of the Division of Biological Sciences actually turned their publications into organs of anti-Michurinist alignment.

Academician Lysenko's report, approved by the Central Committee of the Communist Party of the Soviet Union (Bolsheviks), presents the Soviet Union's scientists, and above all the biologists and representatives of the other branches of natural history, with a series of new questions of principles and requires complete and profound reorganization on the part of scientific institutions of their research work in the domain of biology and of an active conversion of biological science into a powerful weapon for altering living nature to the interests of the construction of a communistic society.

Academician Lysenko's report properly revealed and condemned the scientific incompetence of the idealistic reactionary theories of the followers of Weismannism—Schmalhausen, Dubinin, Zhebrak, Navashin, and others.

Our country, through the efforts of such eminent scientists as I. M. Sechenov, I. I. Mechnikov, K. A. Timiryazev, and A. N. Severtsov, has developed and defended Darwinism from the reactionary attacks on the part of a number of Western European biologists-idealists—Weismann, Morgan, Lotsy, Bateson, and others.

The eminent scientists of our country—V. V. Dokuchayev and V. R. Williams—established the advanced theory of the formation and development of soils. The great Soviet scientist V. R. Williams initiated the productive teaching on the unity of the organism and the soil conditions of its life, and gave the theory on the continuous increase of soil fertility.

The brilliant transformer of nature I. V. Michurin created by his efforts a new epoch in the development of Darwinism. The teaching of I. V. Michurin is founded on the great creative force of Marxist-Leninist philosophy. Michurinist teaching sets for itself the most important task of controlling organic nature; of creating new forms of plants and animals necessary for a socialistic society.

Czarist Russia was incapable of evaluating the significance and transforming force of I. V. Michurin's scientific creative genius.

Michurin was discovered for our people and for advanced science through the genius of Lenin and Stalin. In an epoch of Socialism, Michurin's teaching has proved to be a powerful lever in the matter of the transformation of nature. It has received wide opportunities for its development, and popular acclaim.

If in its old form Darwinism set before itself only the problem of explanation of the evolutionary process, then Michurin's teaching, receiving further development through the works of T. D. Lysenko, has set and solves the problem of controlled alteration of hereditary characteristics of plants and animals, has set and solves the problem of controlling the process of evolution.

T. D. Lysenko and his adherents and students have made an essential contribution to Michurinist biological science, to the goal of the development of socialistic agricultural economy, to the concern of the struggle for abundant yields of agricultural crops and productivity of animal husbandry.

The Praesidium of the Academy of Sciences, USSR, obliges the Division of Biological Sciences, biologists, and all naturalists working in the Academy of Sciences to reorganize their work radically; to assume the leadership in the struggle against idealistic and reactionary teachings in science; against toadyism and servility to foreign pseudoscience. The natural-history scientific institutes of the Academy of Sciences must fight actively for a continual progress of native biological science and, in the first place, for the further development of the teachings established by I. V. Michurin, V. V. Dokuchayev, and V. R. Williams, continued and developed by T. D. Lysenko.

The Praesidium of the Academy of Sciences, USSR, resolves:

1. To release Academician L. A. Orbeli from the duties of Academician-Secretary of the Division of Biological Sciences. To confer temporarily (until election by a general assembly) the duties of Academician-Secretary on Academician A. I. Oparin. To introduce Academician T. D. Lysenko to the staff of the Bureau of the Division of Biological Sciences.

2. To release Academician I. I. Schmalhausen from the duties of Director of the A. N. Severtsov Institute of Morphological Evolution.

3. To abolish the laboratory of cytogenetics in the Institute of Cytology, Histology, and Embryology, directed by Corresponding Member N. P. Dubinin, as being based on antiscientific concepts and as having evidenced its sterility, in the course of a number of years. To close in the same Institute the Laboratory of Plant Cytology, as having a similarly inaccurate and antiscientific trend. To liquidate the Laboratory of Phenogenesis in the A. N. Severtsov Institute of Morphological Evolution.[3]

4. To oblige the Bureau of the Division of Biological Sciences to revise the plans of scientific research problems for the years 1948-50, keeping in mind the exploitation and development of Michurinist teaching and subordination of the scientific research work of the institutions in the biological division to the needs of the country's national economy.

5. To oblige the Editorial-Publishing Council and the Division of Biological Sciences to prepare during 1948-49 an edition of a scientific biography of Michurin in the series "Science Classics."

6. To revise the staffs of the scientific councils of biological institutes and editorial boards of biological journals; to remove from them supporters of Weismannist-Morganist genetics and replace them with representatives of advanced, Michurinist biological science.

7. To commission the Division of History and Philosophy to provide for works on the theoretical generalization of achievements of the Michurinist movement in biology and criticism of the pseudoscientific Weismannist-Morganist movement in the Division's plan of projects.

8. To commission the Bureau of the Division of Biological Sciences to examine the structure, purpose of projects, and composition of the cadres of the Division's scientific institutes. To

[3] The *New York Times* of December 21, 1948, published an air-mail dispatch from C. L. Sulzberger in London, dated December 17. It included the following passage: "The following Soviet scientists have been dismissed from their posts for entertaining reactionary, idealistic views.

"Academician N. P. Dubinin, expert on cytogenetics, whose laboratory has been liquidated; Prof. S. M. Gershenzon, Ukrainian zoölogist and biologist; Academician N. N. Grishko, Ukrainian agricultural scientist; Academician N. G. Kholdony, Ukrainian botanist; Academician L. A. Orbeli, one of the USSR's leading biologists; Academician V. S. Nemchinov, agricultural scientist; Academician I. M. Polyakov, professor of biology; Academician I. I. Schmalhausen, internationally famous professor of 'Darwinism' at Moscow University and zoölogist; Academician Yudintsev, biologist; Academician D. K. Tretyakov, zoölogist; Academician A. R. Zhebrak, geneticist who figured prominently in the Lysenko affair; and Academician M. M. Zavadovski, biologist.

"As a result of this thorough vacuum cleaning process in the realm of the sciences, the Central Committee encouraged the Academy of the Medical Sciences to dismiss the leading researchers, A. G. Gurevich, G. F. Blyakher and C. F. Gawze."

present within one month a plan of reorganization of the A. N. Severtsov Institute of Morphological Evolution and the Institute of Cytology, Histology, and Embryology.

9. The Editorial-Publishing Council to review publishing plans for the purpose of providing for the publication of scientific treatises in the field of Michurinist biology.

10. The Division of Biological Sciences to conduct a general session in October 1948 devoted to the problem of development of Michurinist biological science. The session to be conducted in conjunction with LAAAS, the biological institutes of the republics, branches, and base of the Academy of Sciences, USSR.

11. To commission the Bureau of the Division of Biological Sciences to review the plan of preparation of fellows in the institutes of the Division of Biological Sciences, being influenced in the matter of preparation of the scientific cadres by the interests of development of Michurinist scientific biology.

12. To publish the proceedings of the extended meeting of the Praesidium in the next issue of *Vestnik Akademic Nauk.* [*News of the Academy of Sciences*]

3

SOVIET SCIENCE WILLINGLY AND VOLUNTARILY SERVES THE PEOPLE

(Editorial in *Pravda,* August 27, 1948)

Led by the Party of Lenin and Stalin, the Soviet people are bravely and confidently marching forward toward the flourishing of productive forces and the building up of Communist society. Progressive mankind follows with pleasure the Soviet people's achievement of victory after victory toward this end.

The great strength of our people consists in the fact that they are armed with a revolutionary Marxist-Leninist teaching, the all-conquering force of which is confirmed by the entire experience of history. The policy of the Bolshevik Party and Soviet State in all branches of social, economic, and cultural life of the community is based on strictly scientific principles. Every new step of the Soviet people on the road toward Communism is based on profound scientific knowledge of the perspectives of development.

There is no country of the capitalist world where science is linked with the people. Bourgeois science, which for the sake of profit serves the interest of the capitalist bourgeoisie, is cut off from the people by a Chinese wall and is alien to the people.

Soviet science is strong in that it does not fence itself off from the people, but willingly and voluntarily serves them. There is no country in the world where the results of scientific achievements receive such speedy and mass application as in our country. The bourgeoisie monopolizes scientific achievements for its selfish aims of competitive private enterprise for profit. The scientific discoveries of Soviet scientists, on the basis of the planned Socialist national economy, immediately become the property of all working people.

There is no country in the world where the State shows such care for science and scientists as in our country. The bourgeois states leave science in the power of capitalist monopolies, and condemn scientists to sell themselves to the exploiters or be left to starvation and misery. The Soviet State annually spends thousands of millions of roubles for scientific establishments and improving the living conditions of scientists.

Transform the world in people's interests. Scientific workers of all branches of the leading Soviet science work in an honorable and at the same time responsible way. The great Stalin put before them the task of not only catching up with scientific achievements abroad, but also surpassing them in the near future.

The Soviet scientists are solving this task successfully. This is particularly clear from the August session of the Lenin Academy of Agricultural Sciences, USSR. Academician Lysenko's report on the situation in biological science, and the discussion of this report during the session, showed that Soviet biological science is the most advanced in the world. Michurin's materialist trend in biology is the only scientific one because it is based on the principles of dialectical materialism—the revolutionary transformation of the world in the interests of the people. The Weismann-Morgan idealist trend in biology is pseudoscientific because it is based ultimately on the admission of divine origin in the development of the world—on the passive adaptation of man to permanent and unchanging laws of nature. The struggle of Michurin's followers with those of Weismann is a form of the class ideological struggle of Socialism against capitalism on the international arena, and against the remnants of bourgeois ideology among a section of scientists inside our own country.

As in every struggle, there is no room for a middle course. Michurin's teaching and that of Weismann and Morgan cannot be reconciled.

The results of the session of the Lenin Academy of Agricultural Sciences have demonstrated the complete victory, in theory and in practice, of Michurin's teaching over the Weismann-Morgan trend in biology. The victory of the revolutionary Michurin teaching over the reactionary teaching of Weismann and Morgan is enormously important for the consolidation of the natural-scientific foundations of the Marxist-Leninist outlook, for the ideological upbringing of the progressive Soviet man, and for the practice of Communist construction.

In solving these tasks a particularly big role must be played by the USSR Academy of Sciences. The grim days of 1942, when the Red Army was waging fierce battles against the German-Fascist hordes, are still fresh in our memory. At that time in a telegram to the President of the USSR Academy of Sciences, Stalin wrote:

"I hope that the USSR Academy of Sciences will assume leadership in the innovators' movement in the sphere of science and production, and will become the center of the progressive, Soviet science in the developing struggle against the worst enemy of our people and all other freedom-loving peoples—against German Fascism."

Soviet scientists made an enormous contribution to the world-historic victory of the Soviet people over Fascism.

Great is the role of the Academy of Sciences as the center of progressive Soviet Science in the contemporary conditions of the struggle for Communism. It concentrates the country's best scientific forces.

Unfortunately it must be noted that it does not fulfill its role in all spheres of science. The Praesidium of the Academy of Sciences and the Biology Department Bureau have stood on one side in the sharp ideological struggle against the reactionaries in the sphere of biological science. Moreover, under the flag of an "impartial" attitude toward two opposing trends in biology, the Department of Biological Sciences in actual fact supported in its work the reactionary Weismannist trend, and treated in an offhand manner the supporters of Michurin's teaching.

Struggle against reactionary theories. Consequently the followers of Weismann and Morgan, Schmalhausen, and others have built a firm nest for themselves in the Biological Department. The pages of academic journals have been open to them, where pseudoscientific reactionary views in biology have been propagated. Although in considering the plan of the Academy of Sciences the State Planning Committee of the USSR, for instance, directly pointed to the inadmissibility and harmfulness of the reactionary anti-Michurin trend in the work of biological institutes of the Academy of Sciences, that criticism was ignored. An open struggle was conducted in many biological institutes against the teaching of Michurin, Williams, and Lysenko.

These institutes not only failed to contribute anything to the practice of Socialist agriculture, but inflicted harm on it by their reactionary theories.

How could this situation arise? This occurred above all because the Praesidium of the Academy of Sciences and the Bureau of the Biological Department forgot the most important principle in any science—the Party principle.[4] They pegged themselves to a position of political indifference and "objectivity." The USSR Academy of Sciences forgot the instructions given by V. I. Lenin that "partisanship" is inherent to materialism, and that materialism, whatever phenomena are being considered, must stand openly and directly on the viewpoint of a definite public group. For Soviet scientists this viewpoint is the interests of the working people, and the basis for their world outlook is dialectical materialism.

The recent conference of the Praesidium of the Academy of Sciences has shown that the Academy possesses sufficient creative forces to correct the state of affairs in biological science and to raise high the banner of progressive Soviet science.

Meanwhile, it must be noted that questions of scientific leadership are not limited to biology. The Praesidium of the Academy of Sciences and the Bureau of the Biology Department, having admitted errors in the leadership of biological science, must draw for themselves the necessary conclusions not only in the sphere of biology, but in all other spheres of scientific activity.

In order in actual fact to lead the innovators' movement in the sphere of science and production and to become the true center of progressive Soviet science, the USSR Academy of Sciences must first and foremost carry out precisely and consistently the Leninist-Stalinist principle of "partisanship" in all spheres of science, and steadily enhance the level of Marxist-Leninist knowledge of scientific workers.

The Academy's plans of scientific works must be revised from the viewpoint of fulfillment of the tasks put forward by the Party and the Government in various economic spheres along the road of gradual transition from Socialism to Communism. It is essential to achieve the stage where scientific work carried out by every scientific worker should be considered first and foremost from the angle of serving the needs of the people, who are waging a struggle for the cause of Communism.

[4] This statement is called to the attention of American apologists.

It is necessary resolutely and irrevocably to banish Weismann-Morgan works from the plans, and strengthen the biological institutes by Marxist-Leninist-Michurinist cadres. It is essential to reach a state of affairs where all valuable scientific achievements are immediately put into practice in Socialist economy.

In an address to the great teacher and leader of the Soviet people, I. V. Stalin, the participants of the conference declared: "We promise you, Comrade Stalin, to take up a leading position in the struggle against idealist reactionary teachings, to clear all roads for the unimpeded development of progressive Soviet science for the sake of the great aims of our people, for the sake of the victory of Communism.

"Only marching along these roads can the Academy of Sciences honorably fulfill the tasks put before it by the Party and the Government. There can be no doubt that the thousands of scientific workers of the Academy of Sciences will fulfill the tasks set in the struggle for the prosperity of our progressive science."

4

S. KAFTANOV

Minister of Higher Education in the USSR

(From *Izvestia,* September 8, 1948)

In Support of Michurin's Biological Theory in Higher Institutions of Learning[5]

There are two opposite trends in biological science. One of them is progressive, and materialistic, called Michurin's theory in honor of its founder, the great reformer of nature, Ivan Vladimirovich Michurin; the other is the reactionary, idealistic Weismann's or Mendel-Morgan theory. The founders of this theory were, as it is well known, reactionaries in science: Weismann, Mendel, and Morgan.

In opposition to the Mendel-Morgan trend, Russia developed and, encouraged by the Soviet regime, brought to its full bloom, the great theory of the great modifier of nature I. V. Michurin.

Michurin's materialistic theory has been continually enriched by the works of his followers, with the Academician T. D. Ly-

[5] Reprinted by permission of *Science* (109:90-92. 1942).

senko at their head. This trend in biology has developed into a mighty current which has taken hold of the masses. It inspires millions of collective farmers with faith in the creative power of their efforts and gives them a firm assurance in the realization of new successes in the field of abundancy of farming products.

The Michurinists have proved, not by word, but by demonstration, that it is possible to direct the inborn qualities of animals and plants and to influence the development of animals and plants in a desired manner. Michurin's theory has adopted and developed the best sides of Darwinism. Darwin had explained the evolution of animals and plants from the materialistic point of view. Michurin has developed this knowledge and taught methods of directing the process of producing new species of plants and new species of domestic animals, thus transforming Darwinism into a really practical, creative doctrine.

Michurin's theory is closely bound to another progressive trend in our science—that of Williams' methods of fertilizing the soil, methods that have adopted and developed the best ideas of the Russian classics in the science of agriculture, those of Dokuchayev, Kostychev, and others.

Thanks to the care of the Bolshevist Party and of the Soviet Government, as well as to the personal care of our great leaders, Lenin and Stalin, Michurin's theory has been preserved from oblivion and has become the property of the people. The efforts of Michurin's followers, led by the Academician T. D. Lysenko, have brought it to a new height of achievement. During the last session of the USSR Lenin Academy of Agricultural Sciences, the report of Academician T. D. Lysenko concerning the status of biological science, duly supported by the Central Committee of the All-Russian Communist Party, has been made subject of consideration. This session has become an important factor in the strengthening and further expansion of the progressive biological theory of Michurin. This session has brought to light the opponents of Michurin's doctrine in biology and has dealt a stunning blow to the reactionary Weismann-Morgan theories. This session has proved to be the greatest triumph of the victorious Michurin theory.

The task of the future is now to develop and to spread with the utmost persistence the world's most advanced and most progressive biological theory, that of Michurin. The success of this task will depend upon the system of teaching and of research that will be carried on in higher institutions of learning.

Unfortunately, the theories of Weismann, Mendel, and Morgan, born in foreign countries, have found their supporters in the midst of our biologists.

The first and the most outstanding representative of this pseudoscientific trend in our country was the Leningrad professor Philipchenko and the Moscow professors Koltsov and Serebrovski. They actively professed racist pseudoscience—the so-called eugenics.

The first supporters of the Mendel-Morgan theories—Philipchenko, Koltsov, and others—openly disavowed the materialistic basis of Darwinism; their modern followers and supporters have done their utmost to mask and hide their anti-Darwinistic idealistic opinions and have always carried the banner of Darwinism. Amidst these, the Academician I. I. Schmalhausen has to be named first.

Academician Schmalhausen denies the inheritance of acquired characters. He finds that evolution depends upon mutations which originate directly in the germ cells of the organism and have a quite accidental and indeterminate character, not regulated by the conditions of its life. This idealistic, reactionary theory is fundamentally antagonistic to Darwin's teaching. Nevertheless, Schmalhausen always hid under the banner of Darwinism.

Weismann's reactionary theory has found many adherents and supporters in numerous biological research institutions, as well as in the Universities of Leningrad and Moscow and, especially, in the Timiryazev Academy of Agriculture.

Our homemade Weismannists, such as Academician Schmalhausen, Professors Serebrovski, Zavadovski, Zhebrak, Dubinin, and others, have fought for years the progressive and revolutionary teaching of Michurin's biology.

The laxity of certain ministers and institutions, of the Minister of Higher Education in particular, has encouraged these people and has given them the opportunity of occupying many responsible positions in universities, institutes, etc. Using these opportunities they strongly opposed the Michurinists, hampered their work, sometimes treated them with contempt, and thus did the greatest harm to Soviet science and national agriculture.

It is in the Moscow and Leningrad universities that the bearers of this reactionary-idealistic Weismann's "science" had built their mightiest strongholds.

Thanks to the laxity of the president of Moscow University, as well as of that of the Central Administration of the Universities of the Ministry of Higher Education, efforts have been made in the biological faculty to organize the anti-Michurinists by means of conferences of anti-Darwinistic character with specially selected participants; the Michurin theory was there and then treated with the highest contempt.

Several chairs of the biological faculty of the Leningrad University were also occupied by representatives of formal genetics. Professor Polyansky, vice-president of the University, spoke openly against the Michurin doctrine and its followers. The former vice-president of the biological faculty, Docent Lobashov, actively opposed the appointment of Michurinist biologists.

Many biological chairs at other universities were also occupied by anti-Michurinists—for instance, Professor Polyakov in Kharkov, Professor Gershenson in Kiev, Professor Chetverikov in Gorky. This proves that the Ministry of Higher Education, as well as its Central Administration of Universities, was inefficient in its supervision of biological studies in the universities.

The same is the case in agricultural colleges and even in the most important one, the Timiryazev Academy of Agriculture—several leading biological chairs have been occupied, up to the present, by anti-Michurinists, for example, the anti-Michurinist Professor Zhebrak held the chair of genetics at the Academy. Many other chairs were also held by formal geneticists, as Professor Borisenko and others. Even the Director of the Academy, Professor Nemchinov, shared the convictions of Weismannists and opposed the appointments of Michurinists.

Nevertheless, it is noteworthy that many biological chairs in agricultural colleges are occupied by worthy representatives of the progressive Michurin theory, who worthily carry on the great work of the famous scientist Ivan V. Michurin. Such are: Professor Yakovlev of the Michurin Fruit and Vegetable Institute, Professor Paguibin of the Tashkent Agricultural Institute, Professor Tikhonoff of the Kazan Agricultural Institute.

In other branches of agrobiological science (agriculture, pedology, cultivation of fruits and vetegables) the majority of the chairs in agricultural colleges are occupied by representatives of the progressive Michurin theory.

Strong defects in the teaching of biology can be traced also in medical colleges. The teaching of biology is based in many of these colleges on textbooks permeated with Mendel-Morgan ideas. Many of the chairs of medical institutes are also occupied by supporters of Mendel and Morgan in biology or by persons who, although not actively opposing the Michurin doctrine, are, nevertheless, basing their convictions and pedagogical activities on the spirit of the Weismann-Mendel-Morgan ideas.

Considerable defects in the teaching of biology exist also in teachers' colleges. The Departments of Teachers' Colleges of the Ministry of Higher Education, and the Ministry of Public Education of the Federation of Republics, have not sufficiently opposed the anti-Michurin trend that tended to keep out the Michurinists.

The Ministries of Higher Education, of Agriculture, of Public

Health, and of Public Education have not become aware soon enough of the reactionary character of the anti-Michurinist trend in biology, did not put a stop to the spreading of such reactionary ideas, and did not create a free field for the new generation of Michurin's followers. And there is no doubt that these followers are very numerous.

This has been proved by the session of the Academy of Agriculture when many of the members of the colleges spoke fervently in favor of Michurin's theory.

We must admit that the principal responsibility for the defects of the teaching of biology lies on the Ministry of Higher Education.

We are first of all responsible for not having used enough propaganda for the development of Michurin's doctrine in biology, for having been blind to the dangerous activities of anti-Michurinists in our higher institutions of learning, and for having admitted them to leading roles in many of them. This is the principal cause of the great defects in our teaching of biology, which was in evidence up to the last session of the Lenin Academy of Agriculture.

The first task of the Ministry of Higher Education must now consist in the elimination of defects in the field of biology teaching and in the clearing of the field for Michurin's doctrine. Curricula and programs, textbooks, and methods of teaching and of research must be reëxamined and reorganized as must the entire system of education and training of cadres of scientists and the activities of publishers and of journals. All biological chairs and faculties must be held and supported by qualified Michurinists, capable of developing the progressive Michurin's doctrine.

The success of this reform in the teaching of biology in our colleges will depend most of all upon the right choice of the teaching personnel.

Action along these lines has already been started. The anti-Michurinist Nemchinov, former director of the Timiryazev Academy of Agriculture, has been replaced by Comrade Stoletov, and Academician Lysenko has taken charge of the chair of genetics and selection; Academician Prezent has replaced the anti-Michurinist Yudintsev as dean of the biological faculty of the State University of Moscow, etc. Our present aim is to fill the ranks of the teaching personnel by biologists-Michurinists.

Programs of basic courses must be urgently reëxamined and modified. Many of these programs in the biological sciences are either entirely based on the Mendel-Morgan theories or considerably affected by the revelations of the reactionary biological "science." The reform of those programs should not be delayed even for a single day. This is one of the most urgent aims.

The textbooks of basic courses in biology are quite unsatisfactory. What has been published up to the present in biology for colleges does not guarantee an education based on the progressive Michurin doctrine.

Publishers specializing in biological literature (such as Selkhozgiz, Medgiz, "Soviet Science," the Academy of Science, Uchpedgiz, etc.) have not been aware of their responsibility when publishing theoretical or popular scientific books and other works on biology for colleges.

An illustration of this state of things is given by the publication of a book written by a member of the Academy of Medical Sciences, S. N. Davidenkov, entitled *Evolutionary-Genetic Problems in Neuropathology*. It was published in 1947 with an enthusiastic preface by Academician L. Orbeli. Fully accepting the Mendel-Morgan "theories," the author made an attempt to revise the theory of Engels concerning the humanization of apes under the influence of their working activities.

We must have textbooks based on the progressive Michurin theory.

Schools for higher education will not be able to carry through this reform of teaching in biological sciences if they do not at the same time reconstruct their scientific research work.

Much of this research work, led by partisans of Mendelism-Morganism, had but slight relation to real life dealing with the practice of medicine, agriculture, veterinary science, and animal husbandry. Professor Schmalhausen, for example, holding the chair of Darwinism in the University of Moscow, published volumes of "works" dealing with problems that have nothing to do with the practice of Socialist Construction.

At the University of Kharkov, methods were applied to problems of Darwinism and genetics that had nothing to do with practical needs of life. Docentin Mikhailova occupied herself with "Interspecific Divergence and Crossability in the Genus *Dianthus*." Docent Dubovski had for his objective the elucidation of "The Cytological Basis of the Early Stages of Divergence in Mosquitoes of Different Species and Subspecies." Countless other problems, without any theoretical or practical significance, were likewise pursued, serving sometimes only as evidence for pseudo-scientific conceptions in biology.

Detachment from life, limitations of academic outlooks, practical sterility, such are the qualifications of the scientific work produced by all research carried on by supporters of Mendelism-Morganism. It is necessary to make a sharp change in all scientific research done by our colleges and direct it toward the most active collaboration with requirements of practical life, as well as with the interests of our national economy.

The Party of Lenin-Stalin protects the progressive Soviet science against infiltration of foreign, reactionary influences. The history of our Bolshevist Party serves as an example of a continuous and strenuous fight for a flourishing, progressive science, a science "that has the courage to tear down old traditions, rules, forms, when they prove to be outlived, when they become breaks stopping the onward movement, a science that creates new traditions, new forms, new rules" (Stalin).

The struggle in the field of biology has ended in a complete triumph of Michurin's doctrine, presenting a new stage in the development of materialistic biology.

Thanks to the Bolshevist Party and, personally, to Comrade Stalin, ways for the further triumphant march of the most progressive Michurin biological science are now clear. The scientists of our colleges will apply, from now on, all their energy to the propaganda of Michurin's biology and to the support of undivided rule of Michurin's biological doctrine in our higher institutions of learning.

5

PRAESIDIUM OF THE ACADEMY OF MEDICAL SCIENCES, USSR

Letter to Comrade Iosif Vissarionovich Stalin
(From *Pravda,* September 15, 1948)

Dear Iosif Vissarionovich:

The grand session of the Praesidium of the Academy of Medical Sciences, USSR, devoted to a discussion of the status and mission of medical science in relation to the directive on biology, addresses you, our beloved leader and teacher, with profound love and gratitude for the constant concern and aid which you have extended to progressive Soviet science as its great leader and friend.

All our best achievements and the entire progressive trend of our science, dear Iosif Vissarionovich, are due to you.

Your wise instructions on the development of science in the

people's interest, on the close tie between theory and practice, on the irreconcilable struggle with all types of reactionary, idealistic theories and trends in science and ideology, are for us a daily guide to action.

You have given great support to progressive Soviet scientists and innovators, headed by Academician T. D. Lysenko, in their long struggle against hostile idealistic tendencies of the Weismann-Morgan type, and to the progressive Michurin teaching in biology.

Academician T. D. Lysenko's report on "The State of Biological Science," approved by the Central Committee of the ACP, has historic significance for the development of the whole science of biology.

Vladimir Ilich Lenin and you, his great successor, have discovered Michurin, the brilliant scientist of the Russian people, and have set his supreme scientific labor for the welfare of the people as an example of what the scholarly Soviet innovator should do.

Thanks to the concern of the Party, the Government, and you personally, our dear teacher, the progressive Michurin teaching in agricultural biology has become a guide to the action of millions of agricultural workers and has contributed to their remarkable practical achievements.

Only the ideological defeat of the idealistic, bourgeois tendencies of Weismann and Morgan, which have found adherents among some of our Soviet scientists, with their abject servility before bourgeois science, has insured the success of our progressive Michurin science.

But Weismann-Morganism, which is a fake bourgeois science, has also had a baneful influence on the development of medical science.

Among our scientists there are several devotees of this reactionary trend, which has inculcated the idealistic teachings of Mendel, Weismann, and Morgan in the field of medicine.

To a great degree we are responsible for this because of our tolerant attitude toward these idealistic views in medical and biological science. As a result of the weakness of Bolshevik criticism and self-criticism in our academic circle, the followers of Weismann and Morgan have had an opportunity to spread their harmful ideas in medical science to the detriment of the development of the fruitful ideas of such founders of Russian medical science as Pirogov, Sechenov, Pavlov, and Mechnikov.

In the work of the V. I. Lenin Academy of Agricultural Science, Soviet medical science has received a powerful new impetus to its further development.

The Academy of Medical Sciences, USSR, with all its scientific collectives, is faced with the urgent task of eradicating the reactionary doctrines of Weismann-Morganism from medical science.

Armed with the teachings of Marx, Engels, Lenin, and Stalin, we will devote all our efforts to solve the urgent problems of medicine and public health and to fight for the health of our great Soviet people.

We promise you, our dear leader, to rectify in the shortest possible time the mistakes we have permitted to occur, and to reconstruct our scientific work in the spirit of the directives issued by the great Party of Lenin and Stalin. We promise you to master fully the great Michurin teaching and to utilize it for the further development of Soviet medicine.

We will fight for the Bolshevik Party spirit in medicine and public health and will eradicate hostile bourgeois ideology and servility before foreign isms in our midst. Promoting criticism and self-criticism in scientific work, we will labor indefatigably to fulfill your wise instructions and to develop medical science for the welfare of our people, the builders of Communism.

THE PRAESIDIUM OF THE ACADEMY OF MEDICAL SCIENCES, USSR

Pravda, September 15, 1948

XII

Reverberations Abroad

When the stenographic reports of the Sessions of the Lenin Academy were published in *Pravda* (August 4-12, 1948), the news spread quickly throughout the scientific world. Geneticists who had followed the attacks on their science in 1936 were not too surprised, although the course of events in Russia had led many of them to believe that Lysenko was a passing fad, that his pretensions would soon be exposed and that he would quietly disappear in the conventional Russian manner. It was only when his *Concluding Remarks* were published that they learned that Lysenko had entered into the Apostolic Succession and consequently could not be "in error."

Scientists in other disciplines and the public at large showed great interest in the destruction of genetics. Literally hundreds of newspaper and magazine articles appeared on the subject. These are in a fugitive form, and unfortunately most of them will be lost. The Supplementary Bibliography at the end of this chapter lists a number of the more important of these items, and Morris C. Leikind is preparing a more complete bibliography on the subject for the *Journal of Heredity*. Any adequate collection of this material would constitute a volume in itself, and it should be collected and published. Here attention can be called only to its existence. Indeed the interest in the subject was so great that the British Broadcasting Corporation staged a radio debate which was published later in the *Listener*.

The communist press outside of Russia naturally supported both the scientific claims of Lysenko and the acts of the Soviet authorities. It is particularly important that this portion of the material be put into a permanent form, for it would be of major interest to logicians, psychologists, and historians of science. Any adequate history of our times needs this source material. Perhaps nothing else shows so well the general intellectual deterioration of the communist's position. When genetics was first attacked in Russia our Party liners denied that science was being subordinated to political dogma and that scientists were being persecuted. Events, however, have moved rapidly, and the connection between science and politics in Russia must now be admitted (p. 311), although at the same time it is being defended as a new and improved something or other. The reader is urged to compare the articles on the subject in the earlier issues of *Science and Society, Philosophy of Science, Scientific Monthy,* etc., with some of the more recent defenses of Lysenko and the Central Committee. The nadir was probably reached in a column by Milton Howard in the *Daily Worker* of January 16, 1949.

This chapter will contain but four items. First, a statement of the Governing Board of the American Institute of Biological Sciences; second, the resignation of Professor H. J. Muller from the Academy of Sciences, USSR; third, the reply of the Praesidium of the Academy of Sciences to Professor Muller's resignation; and fourth, the resignation of Sir Henry Dale from the same Academy.

1

THE GOVERNING BOARD OF THE AMERICAN INSTITUTE OF BIOLOGICAL SCIENCES[1]

Statement

For more than a decade, biological scientists and particularly geneticists and cytologists in the USSR have been attacked by so-called "Michurinists," led by T. D. Lysenko, now a high government official and a public figure. Lysenko and his followers have declared the principal attainments of genetics and cytology, including Mendel's laws, to be invalid. This has been done in a manner which shows clearly that Lysenko is either unfamiliar with, or else is willfully ignoring, the basic facts and the methods of investigation of the sciences which he presumes to negate. On the other hand, Lysenko and his adherents have claimed successful experiments with higher organisms demonstrating directed hereditary changes of a useful kind, by means of adaptive responses that later were inherited. Such phenomena would have been of great theoretical and practical value if confirmable. However, outside Lysenko's group in the USSR, such confirmation has proved impossible.

[1] Reprinted by permission of *Science* 110:124-25. 1949.

The necessity for clarifying the situation becomes all the greater because Russian spokesmen, such as I. I. Prezent and S. Kaftanov, quote from the works of Western geneticists in support of their views. This Communist party line has even penetrated in subtle ways into reputable weekly and daily journals in France, England, and the United States. The opinion is consequently spreading that modern genetic researches in the West support the official Communist views on heredity. Nothing could be farther from the truth. Authors quoted by the Russians have strongly denied the validity of drawing such conclusions from their studies. In no case has their work discredited or contradicted the firmly established validity of the gene. They resent having their papers cited as leading to such a discrediting, for this is a manifest reversal of their data and of the intent of their statements.

The opinions and claims of Lysenko and his followers have become a matter of especially serious concern to scientists everywhere because the government of the USSR has not only approved and supported the Lysenko group, but has also condemned and suppressed those biologists in the USSR who have disagreed with Lysenko and who have tried to continue their research in the fields of genetics, cytology, and related sciences. As reported in recent months by the world press, and published in the official newspapers of the USSR, the views of Lysenko have been endorsed by the government organs directing scientific research in that country—among them the Academy of Sciences of the USSR, the Academy of Agricultural Sciences, and the Academy of Medicine. More serious still, geneticists, cytologists, and evolutionists as eminent in their fields and as well known to their colleagues all over the world as Dubinin, Schmalhausen, Zavadovsky, and others have been removed from their positions, deprived of their laboratories, or led to make shameful declarations of their supposed acceptance of Lysenko's view. Finally, the temper of this supposedly scientific controversy may be appreciated by the pronouncements of S. Kaftanov, Minister of Higher Education, to the effect that all anti-Lysenko doctrine must be systematically rooted out of the schools, universities, research institutes, and publishing houses.

It may be left to the judgment of scientists, friends of science, and all fair-minded people to arrive at their own conclusions regarding the propriety of governments and political parties not only deciding a supposedly scientific controversy in favor of one and against another theory, but also dismissing scientists and depriving them of the means of conducting their research, and too often of their lives, because of their adherence to a scientific theory accepted everywhere on this side of the iron curtain. As representatives of American scientific societies devoted to furtherance of research and study in genetics, we feel it our duty to state that the contention raised by Lysenko and his "Michurinists" against genetics does not represent a controversy of two opposing schools of scientific thought. It is in reality a conflict between politics and science. Today the condemned science happens to be genetics. Indeed, the conflict has already spread to other biological fields, and eminent physiologists, embryologists, microbiologists, and others are now being dismissed in the USSR. Tomorrow still other sciences may be proscribed.

The progress of science has always depended upon free inquiry. The inheritance of acquired characteristics, and other doctrines that the Russians now set forth as the official party line, have had their proponents in America; some nongeneticists still hold to these ancient opinions. Nevertheless, they are allowed to investigate or philosophize, and they have a hearing. In Russia, on the other hand, geneticists are being rooted out as dangerous, bourgeois, reactionary, idealist, fascist, regardless of their political views, simply because they, like geneticists everywhere else in the world, know and accept the facts of experimental breeding and microscopic observation which Russian politics has branded false. It is of the utmost importance for the preservation of free inquiry in that part of the world where it still exists that these facts be known and fully appreciated.

The Governing Board of the American Institute of Biological Sciences, an organization representing American societies in numerous fields of biology, is issuing the present statement after consultation with the executive committees of those societies in its organization which deal more particularly with the matters here at issue—namely, the Genetics Society of America and the American Society of Human Genetics. We would sum up our positions in the following propositions:

1. In our opinion the conclusions of Lysenko and his group regarding the inheritance of adaptive responses in higher organisms have no support in scientific fact.

2. Genetic researches definitely support the reality of the gene and the validity of Mendel's laws. They do not support the official Communist claim that Mendelian heredity is an illusion, and any attempts on the part of Russian proponents of the Lysenko doctrines to bolster their case by citations from the works or conclusions of Western scientists are gross distortions of the meaning and intent of these scientists.

3. We condemn the action of the Soviet government in presuming to banish a firmly established science from its schools,

publishing houses, and research laboratories, and in persecuting scientists because their field of inquiry is distasteful to the government.

E. G. Butler
T. C. Byerly
F. P. Cullinan
W. O. Fenn
R. E. Cleland
Executive Committee

Governing Board
American Institute of Biological Sciences

2

RESIGNATION OF PROFESSOR H. J. MULLER FROM THE ACADEMY OF SCIENCES, USSR

[The following letter, dated September 24, 1948, was sent by H. J. Muller, of Indiana University, Nobel Prize winner and past president of the Genetics Society of America, to the President, the Secretary, and the Membership of the Academy of Sciences of the USSR:]

In February 1933 the Academy of Sciences of the USSR sent me a diploma, signed by its venerable President, Karpinsky, and its Secretary, Volgin, stating that I had been elected a "Corresponding Member." In accepting this election, I realized that it was a signal honor, inasmuch as your Academy had a long and most distinguished tradition of scientific achievement and integrity, and was still maintaining its high standards and, in fact, greatly expanding its valuable work. Although for nearly a decade I have not been sent your publications, I must presume that I am still on your rolls, since I have received no information to the contrary.

The deep esteem in which I have held your organization in the past makes it the more painful to me to inform you that I now find it necessary to sever completely my connection with you. The occasion for my doing so is the recently reported series of actions of your Praesidium in dropping, presumably for their adherence to genetics, such notable scientists as your most eminent physiologist, Orbeli, and your most eminent student of morphogenesis, Schmalhausen, in abolishing the Laboratory of Cytogenetics of your most eminent remaining geneticist, Dubinin, in announcing your support of the charlatan, Lysenko, whom some years ago you had stooped to take into your membership, and in repudiating, at his insistence, the principles of genetics. These disgraceful actions show clearly that the leaders of your Academy are no longer conducting themselves as scientists, but are misusing their positions to destroy science for narrow political purposes, even as did many of those who posed as scientists in Germany under the domination of the Nazis. In both cases the attempt was made to set up a politically directed "science," separated from that of the world in general, in contravention of the fact that true science can know no national boundaries but, as emphasized at the recent meeting of the American Association for the Advancement of Science, is built up by the combined efforts of conscientiously and objectively working investigators the world over.

In Germany too it was the field of genetics, that of my own specialization, which was subjected to the greatest perversion, as I pointed out in publications and lectures gotten out both shortly before and during several years after the Nazi coup. And in the USSR the prescientific obscurantism of Lysenko, supported by the so-called "dialectical materialism" represented by Prezent, with their faith in the inheritance of acquired characters, must lead inevitably, and indeed by the admission of some of their adherents, to the same dangerous Fascistic conclusion as that of the Nazis: that the economically less advanced peoples and classes of the world have become actually inferior in their heredity. The Nazis would have the allegedly lower genetic status a cause, while the Lysenkoists would have it an effect, of the lower opportunity of the less fortunate groups for mental and physical development, but in either case a vicious circle is arrived at, which objective geneticists do not concede. Objective geneticists, on the contrary, having established the existence of a separate material of heredity, which is not influenced in any corresponding way by modifications of the phenotype, or bodily characteristics of organisms, recognize the fallacy of judging the hereditary endowments either of individuals or of whole groups simply by outward appearances. Especially is this the case when, as with human mental traits, there are very variable environmental influences, such as differences in tradition, education, nutrition, etc., which have pronounced and systematic effects upon the development of these characters.

In truth, genetics is so fundamental and so central to all fields of biological science, and even of social science and philosophy, that the excision of its established principles from the body of science as a whole cannot but result in the eventual debilitation and falsification of our understanding of things in general. Even the physical sciences must in the end be adversely affected by the admission of the naïve and archaic mysticism of Lysenko, Prezent, and their group into the vacuum left by the removal of genetics, for processes must then be invoked which are contradictory to the workings of matter.

Under the circumstances above set forth, no self-respecting scientist, and more especially no geneticist, if he still retains his freedom of choice, can consent to have his name appear on your list. For this reason I hereby renounce my membership in your Academy. I do so, however, with the ardent hope that I may yet live to see the day when your Academy can begin to resume its place among truly scientific bodies.

The importance of the matters here at issue—including that of the authoritarian control of science by politicians—is in my opinion so profound that I am making this letter public.

3

A REPLY TO PROFESSOR H. J. MULLER

By the Praesidium of the Academy of Sciences, USSR
Pravda, December 14, 1948

(Quoted by permission from *Soviet Press Translations* 4:48-49)

Professor H. J. Muller, of Indiana University, U. S. A., sent the President of the Academy of Sciences, USSR, a letter, also published in the American press, in which he expresses disagreement with recent decisions of the Praesidium of the Academy of Sciences, USSR, on questions relating to biological science and announces his resignation as a member of the Academy.

The contents of this letter bear witness that Professor Muller, in defining his position on scientific questions, is not guided by the interests of science, and by the interests of truth.

He attempts to accuse the Academy of Sciences of the Soviet Union to the effect that, in the dispute between the followers of the views of Weismann-Morgan and the adherents of the genuine Darwinian theory, which has been elevated to a new and loftier level in the works of the outstanding Russian scholar

Michurin, it has definitely adopted the position of the Michurin biology. In this connection, Muller is far from acting as behooves an objective investigator, examining the reasons and arguments advanced by the opponents of the Weismann school. He has ignored the well-known dictum of Pavlov that "facts are the air of the scholar," and preferred to reject the analysis of numerous new experimental data obtained by Soviet investigators. Such conduct can hardly be regarded as a sample of scientific courage and objectivity.

The Academy of Sciences, USSR, in defining its position, proceeded on the basis of experimental work, which clearly showed the erroneous concepts of Weismann-Morgan and the triumph of the ideas of Michurin.[2] Michurin and his successor, Academician T. D. Lysenko, rejected the teaching of a special "hereditary substance," independent of the rest of the organism and of its conditions of life. Having discovered the link between heredity and the living conditions of organisms, the Michurin biology advanced concrete methods for a conscious, controlling influence over the organic world, and pointed out the way to transform the nature of organisms in the direction necessary to man. Michurin biology rejects the power of accidents, of blind elemental forces, in one of the most important spheres of human activity—in the process of producing new types of plants and new breeds of animals.

The Michurin teaching is a further stage in the development of Darwinism. Having imbibed the huge wealth of ideas of Darwin, as well as the leading, progressive views of other biologists, including Lamarck, having further cast aside their fallacies, the teaching of Michurin and his followers has raised biology to a new and higher level.

Profound indignation is aroused by the assertion of Muller to the effect that Michurin biology allegedly leads to racist conclusions, inasmuch as it allegedly follows that the living conditions of culturally backward peoples must determine hereditary incapability of adopting a higher culture. This nonsense has nothing in common with Michurin science.

Soviet scholars categorically reject the attempts to apply biological laws to social life, including the conclusions of Michurin biology. The development of society is subject not to biological but to higher social laws. Every attempt to extend the laws of the animal world to mankind means an attempt to lower man to the level of a beast.[3]

[2] No data whatever is cited to substantiate this claim.

[3] We Americans can top even this. During the famous Scopes Trial in Tennessee, William Jennings Bryan denied that man was a mammal.

Not Michurin biology, but Morgan genetics, is being used in every way possible by reactionaries to substantiate racial theories. It was so in fascist Germany and it is being done now in America. And Professor Muller, President of the Genetics Society of America, some time ago came out against eugenics, and is now a propagandist of man-breeding, and has joined forces with the most avowed racists and reactionaries in science.

Professor Muller expresses his disdain for "so-called dialectical materialism," which is inseparably bound up with the Michurin teaching. It should be recalled here that in 1934 Muller expressed other views and called upon geneticists "to rid themselves of antimaterialistic and antidialectical propensities." We do not believe in the sincerity of these words. However, Soviet science, not in words but in reality, is mastering the method of dialectical materialism, for this method is the most reliable weapon of scientific research. Regarding everything as in motion, in development, in interrelationship, dialectical materialism rejects the existence of certain unchangeable substances, such as an "hereditary substance," affirms the closest connection between the organism and the environment. Perhaps it is precisely this which Muller does not like.

Muller declares that in its decision on the biological question the Academy of Sciences was motivated by political goals, that science in the Soviet Union is subject to politics.

We, the Soviet scientists, are convinced that the entire experience of history teaches that there does not exist and cannot exist in the world a science divorced from politics. The fundamental question is with what kind of politics science is connected, whose interests it serves—the interests of the people or the interests of the exploiters.

Soviet science serves the interests of the common people; it is proud of its connection with the policy of the Soviet State, which has no other aims but the improvement of the welfare of the workers, and the strengthening of peace and the progress of democracy. The close, creative collaboration of Soviet biological science with the policy of the State is demonstrated more clearly than anything else in the practical work of increasing the yield in agricultural crops, livestock-breeding, and in realizing plans unprecedented in scope and marvelous in concept against drought by planting thousands of kilometers of forests.

Soviet scientists regard with feelings of the warmest sympathy those foreign scientists who honestly and sincerely strive to serve the interests of peace, progress, and democracy. However, it is impossible to shut one's eyes to the fact that in capitalistic countries the bourgeoisie subordinates scientific research to its interests, in pursuit of the goals of profit, oppression of the working masses, and suppression of democracy. At the present time, it is common knowledge that there is taking place in the United States the militarization of science, the subjection of science to the usurping, aggressive plans of American imperialism.

One wonders why Professor Muller does not come out against the utilization of scientific achievements by American imperialists for the purpose of the mass destruction of people and cultural treasures,[4] but rather against Michurin biological science, which pursues the task of the speediest possible advancement of the people's welfare.

The successes of the Soviet people in building a new Communist society, the flourishing of culture, science, and art in the USSR, depend on the ever growing authority of and profound respect for the Soviet land on the part of all the progressive people of the world. On the other hand, these successes evoke malice and hatred on the part of the enemies of genuine science and progress—the imperialists and their hirelings.

Professor Muller was once known as a progressive scientist. This is a very uncomfortable position in present-day America. Having come out against the Soviet Union and its science, Muller has won the enthusiasm and recognition of all the reactionary forces of the United States.

The Academy of Sciences, USSR, without any feeling of regret parts with its former member, who betrayed the interests of real science and openly passed over into the camp of the enemies of progress and science, peace and democracy.

PRAESIDIUM OF THE ACADEMY OF SCIENCES, USSR

[4] In his article "Back to Barbarism—Scientifically" (*Saturday Review of Literature,* Dec. 11, 1948), Professor Muller does call attention to certain dangers that he believes threaten science in the United States.

4

RESIGNATION FROM THE ACADEMY OF SCIENCES OF THE USSR

By SIR HENRY H. DALE, O.M., G.B.E., F.R.S.

Formerly President of the Royal Society, and of the British Association for the Advancement of Science

London
November 22nd, 1948

To The President of the Academy
of Sciences of the U.S.S.R.,
Moscow.

Mr. President,

I have come to the conclusion that I ought to resign the Honorary Membership of the Academy of Sciences of the U.S.S.R., to which I was elected in May, 1942.

When your distinguished predecessor, Academician V. L. Komarov, wrote to inform me of the honour thus conferred upon me, his letter made reference to the fact that I was then President of the Royal Society of London. I believe that many British scientists, indeed, recognised and welcomed my election at that time as a symbol of a community of purpose between the scientists of our two nations, which had then been engaged together for nearly a year in the war against Hitler's Germany in defending, as we believed, the freedom of Science, as of all man's proper activities, from the threat of an aggressive tyranny.

In that same year, 1942, the Royal Society of London elected Nicholas Ivanovitch Vavilov to be one of its fifty Foreign Members. With Lenin's support and encouragement he had been able, as first Director of the Lenin Academy of Genetics, to initiate and promote a rapidly-growing contribution, by research in the U.S.S.R., to that world-wide advance in the science of Genetics which had followed the recognition of Mendel's discoveries. His use of these opportunities was reputed to have been of great benefit to Agriculture in the Soviet Union; we desired to honour it as a great contribution to Science for the whole world.

It had been reported in Britain, however, already in 1942, that N. I. Vavilov had somehow fallen from favour with those who came after Lenin, though the cause of his trouble was still unknown; we might have supposed it to be political, or otherwise irrelevant to his scientific achievement. Not till 1945 did the Royal Society discover that he had been dismissed from his position, had disappeared with a number of his co-workers in Genetics, and had died at some unknown date between 1941 and 1943. Repeated inquiries, addressed to your Academy by the Royal Society through all available channels, asking only the date and the place of his death, received no reply of any kind. I understand that the Royal Society has not yet been officially informed whether this distinguished Russian scientist was still alive at the time of his election to its Foreign Membership.

More recent events, of which full reports have come to hand, have made it clear what has happened. The late N. I. Vavilov has been replaced by T. D. Lysenko, the advocate of a doctrine of evolution which, in effect, denies all the progress made by research in that field since Lamarck's speculations appeared early in the 19th century. Though Darwin's work is still formally acknowledged in the U.S.S.R., his essential discovery is now to be rejected there. The whole great fabric of exact knowledge, still growing at the hands of those who have followed Mendel, Bateson and Morgan, is to be repudiated and denounced; and the last few, who were still contributing to it in the U.S.S.R., have now been deprived of position and opportunity.

This is not the result of an honest and open conflict of scientific opinions; Lysenko's own claims and statements make it clear that his dogma has been established and enforced by the Central Committee of the Communist Party, as conforming to the political philosophy of Marx and Lenin. Many of us, Mr. President, have been proud to think that there were no political frontiers or national varieties in a Science common to all the world; but this is now to be separated from 'Soviet Science' and repudiated as 'bourgeois' and 'capitalistic.'

Decrees which the Presidium of your Academy has issued, on August 27th of this year, give effect only too clearly to this political tyranny. My old and honoured friend, Academician L. Orbeli, distinguished neurophysiologist of the school of your great Pavlov, is dismissed from his Secretaryship of your Academy's Department of Biological Sciences, because he has failed to anticipate your Decrees, in their restriction of all research and teaching in Genetics in the U.S.S.R. to this politically imposed orthodoxy. It remains to be seen whether such compliance with dogma is to be exacted in other departments of Science. So far, we know only that the Genetics encouraged by Lenin is now prohibited as alien to his political philosophy.

Since Galileo was driven by threats to his historic denial, there have been many attempts to suppress or to mutilate scientific truth in the interests of some extraneous creed, but none has had a lasting success; Hitler's was the most recent failure. Believing, Mr. President, that you and your colleagues must be acting under a like coercion, I can only offer you my respectful sympathy. For my own part, being free to choose, I believe that I should do disservice even to my scientific colleagues in the U.S.S.R., if I were to retain an association in which I might appear to condone the actions by which your Academy, under whatever compulsion, is now responsible for such a terrible injury to the freedom and the integrity of Science.

With deep regret, I must ask you to accept my resignation.

Yours faithfully,
HENRY H. DALE.

Bibliography

This bibliography does not list papers quoted in whole in the preceding pages, nor does it contain the papers cited therein which have only an incidental connection with the destruction of genetics. The action of the Central Committee of the Communist Party in approving Lysenko's position has caused a world-wide repercussion. Lysenko, himself, has all of the news value of the man who bit the dog. The result has been an enormous fugitive literature both in the newspapers and in certain political periodicals, which are probably ephemeral. Some of these items, which are individually trivial but whose mass and number help to give a picture of our contemporary partisan thinking, are included. Most of the following papers, on the other hand, are intrinsically important and are extremely useful for anyone who wants to know more about the problems involved in the recent emergence of a political control of science.

Aragon, Louis: "Storm over Lysenko," *Masses and Mainstream* 2:22-38. 1949.

Arnot, R. Page: "Scientists in Livery," *Labour Monthly* 31:55-59. 1949.

Ashby, Eric: "Genetics in the USSR," *Nature* 158:285-87. 1946.

———: *Scientist in Russia. New York* (Penguin Books), 1947.

Azcoaga, J. E.: "Sobre la Genética en la Unión Soviética," *Ciencia e Investigacion* 5:156-60. 1949.

Baker, John R.: *Science and the Planned State*. London, 1945.

———: "The Soviet Genetics Controversy," *Time and Tide* 29:1297. 1948. Society for Freedom in Science Occasional Pamphlets No. 9. 1949.

———, and A. G. Tansley: "The Course of the Controversy on Freedom in Science," *Nature* 158:574-76. 1946.

Beale, G. H.: "Timiriazev: Founder of Soviet Genetics," *Nature* 158:51-53. 1947.

Cannon, Walter B., *et al.*: *Science in Soviet Russia*. Lancaster, Penna., 1944.

Chevalier, Aug.: "La Polémique des Biologistes mitchuriniens et mendélo-morganiens en URSS," *Rev. Intern Bot., Appl.* No. 315-16:1-17. 1949.

Cook, R. C.: "Lysenko's Marxist Genetics," *Journal of Heredity* 40:169-202. 1949.

———: "Walpurgis Week in the Soviet Union," *Scientific Monthly* 68:367-72. 1949.

Crane, M. B.: "Lysenko's Experiments," *Bulletin of the Atomic Scientists* 5:147-79. 1949.

Darlington, C. D.: "A Revolution in Soviet Science," *Journal of Heredity* 38:143-48. 1947.

———: "The Lysenko Controversy," *The Listener,* Dec. 9, 1948.

Davies, R. G.: "Genetics in the USSR," *Modern Quarterly* 2:336-46. 1947.

Deutsch, Albert: "Wallace Gives Views on Lysenko Dispute," *Daily Compass,* July 12, 1949.

———: "Some Reflections on the Great Lysenko Debate," *Daily Compass,* July 12, 1949.

Dobzhansky, Th.: "Lysenko's Genetics," *Journal of Heredity* 37:5-9. 1946.

———: "The Suppression of a Science," *Bulletin of the Atomic Scientists* 5:144-46. 1949.

Dubinin, N. P.: "Work of Soviet Biologists," *Science* 105:109-12. 1947.

Dunn, L. C.: Review of Lysenko's *Heredity and Its Variability, Science* 103:180-81. 1946.

———: "Science and Politics in Russia," *Bulletin of the Atomic Scientists* 12:368. 1948.

———: "Motives for the Purge," *Bulletin of the Atomic Scientists* 5:142-43. 1949.

Espinasse, P. G.: "Genetics in the USSR," *Nature* 148:739-43. 1941.

Feuer, Lewis S.: "Dialectical Materialism and Soviet Science," *Philosophy of Science* 16:105-24. 1949.

Fisher, R. A.: "The Lysenko Controversy," *The Listener,* Dec. 9, 1948.

Frankel, O. H.: "The Lysenko Controversy," *The Listener,* Dec. 9, 1948.

Friedman, Bernard: "Lysenko's Contribution to Biology," *Soviet Russia Today,* Jan. 1949.

———: "Revolution in Genetics," *Masses and Mainstream* 2:40-42. 1949.

Fyfe, J. L.: "The Soviet Genetics Controversy," *Modern Quarterly* 2:347-51. 1947.

Glass, Bentley: Review of *The Science of Biology Today* by T. D. Lysenko. *Science* 109:404-5. 1949.

Glushchenko, I.: "Reactionary Genetics in the Service of Imperialism," *Pravda,* April 5, 1949. *Soviet Press Translations* 4:339-41. 1949.

Goldschmidt, Richard: Review of Lysenko's *Heredity and Its Variability, Physiological Zoology* 19:332-34. 1946.

———: "Research and Politics," *Science* 109:219-27. 1949.

———: "Research and Politics," *Bulletin of the Atomic Scientists* 5:150-55. 1949.

Haldane, J. B. S.: "Lysenko and Genetics," *Science and Society* 4:433-37. 1940.

Haldane, J. B. S.: "The Lysenko Controversy," *The Listener*, Dec. 9, 1948.
Harland, S. C.: "The Lysenko Controversy," *The Listener*, Dec. 9, 1948.
Howard, Milton: Column in *Daily Worker*, Jan. 16, 1949.
Hruby, Karel: "Theoretcǩy zóklad Mithurinske genetiky," *Vesmír*, p. 82. 1949. English summary.
Hudson, P. S., and R. H. Richens: "The New Genetics in the Soviet Union," *Imperial Bureau of Plant Breeding and Genetics*. Cambridge, 1946.
Huxley, Julian: "Science in the USSR: Evolutionary Biology and Related Subjects," *Nature* 156:254-56. 1945.
———: "Soviet Genetics: the Real Issue," *Nature* 163:935-42; 974-82. 1949.
Kartman, L.: "Soviet Genetics and the 'Autonomy of Science,' " *Scientific Monthly* 61:67-70. 1945.
Krasnov, Mikhail: "Michurin Altered Plant Heredity and Upset Old Biology," *USSR Information Bulletin* 9:184-85. 1949.
Langmuir, I.: "Science and Incentives in Russia," *Scientific Monthly* 63:85-92. 1946.
Laptev, I.: "The Triumph of Michurin Biological Science." Review of *Proceedings of the Lenin Academy of Agriculture*, July 31 to Aug. 7, 1948. *Pravda*, Sept. 11, 1948. *Soviet Press Translations* 3:686-94. 1948.
Leikind, Morris C.: "The Genetics Controversy in the USSR," *Journal of Heredity* 40:203-8. 1949.
Lenin Academy of Agricultural Sciences, USSR: *Situation in the Biological Sciences. Proceedings*. Moscow (Foreign Language Publishing House), 1949.
Lenin Academy of Agriculture, Session of the: "Science on the Position of Biological Science," *Plant Breeding Abstracts* 18:642-44. 1948.
Levy, Jeanne: "The Genetics Controversy; II. Lysenko and the Issues in Genetics," *Science and Society* 13:55-78. 1949.
Lewis, John: "A Footnote on the Soviet Genetics Controversy," *Modern Quarterly* 2:352-56. 1947.
Lysenko, Trofim: *Heredity and Its Variability*. New York, 1946.
———: *The Science of Biology Today*. New York, 1949.
Mather, K.: "Genetics and the Russian Controversy," *Nature* 149:427-30. 1942.
Muller, H. J.: "The Destruction of Science in the USSR," *Saturday Review of Literature*, Dec. 4, 1948.
———: "Back to Barbarism Scientifically," *Saturday Review of Literature*, Dec. 11, 1948.
———: "The Crushing of Genetics in the USSR," *Bulletin of the Atomic Scientists* 12:369-71. 1948.
———: "It Still Isn't a Science," *Saturday Review of Literature*, Apr. 16, 1949.

Nicolaevsky, Boris I: "Palace Revolution in the Kremlin," *The New Leader* 32:(12):8: March 19, 1949.

Oparin, A. I.: "Science and the Struggle for Peace," *Soviet Russia Today*, May 1949.

Pauling, Linus, *et al.*: "The Spitzer Case," *Chemical and Engineering News* 27:935-36. 1949.

Pincus, J. W.: "The Genetic Front in the USSR," *Journal of Heredity* 31:165-68. 1940.

Polyani, Michael: "The Autonomy of Science," *Scientific Monthly* 60:141-50. 1945.

Prasalov, L. I.: "Tasks of the Journal *Pedology* in the Light of the Resolutions of the All Union Academy of Agriculture and the Presidium of the Academy of Sciences, USSR," *Soils and Fertilizers* 12:69-71. 1949.

Prenant, Marcel: "The Genetics Controversy; I. The General Issues," *Science and Society* 13:50-54. 1949.

Rabinowitz, Eugene: "The Purge of Geneticists in the Soviet Union," *Bulletin of the Atomic Scientists* 5:130. 1949.

———: "History of the Genetics Conflict," *Bulletin of the Atomic Scientists* 5:131-40. 1949.

Sadosky, Manuel: "A Própositio de la Genética en la URSS," *Cienca e Investigacion* 5:160-62. 1949.

Samachson, Joseph: "In Defense of Lysenko," *Scientific Monthly* 69:209-10. 1949.

Sax, Karl: "Soviet Science and Political Philosophy," *Scientific Monthly* 65:43-47. 1947.

———: "Genetics and Agriculture," *Bulletin of the Atomic Scientists* 5:143. 1949.

Schechter, Amy: "Citrus Moves North," *Soviet Russia Today*, May 1949.

Shaw, G. Bernard: "Behind the Lysenko Controversy," *Saturday Review of Literature*, April 16, 1949.

Somerville, John: "Soviet Science and Dialectical Materialism, *Philosophy of Science* 12:23-29. 1945.

Spitzer, Ralph W.: "Spitzer Answers," *Chemical and Engineering News* 27:909. 1949.

Stark, Morton: "The Stockholm Congress," *Journal of Heredity* 39:219-21. 1948.

Stern, Bernhard J.: "Genetics Teaching and Lysenko," *Science and Society* 13:136-49. 1949.

Stone, Peter: "Soviet Science Is Changing Heredity," *Daily Worker*, Dec. 26, 1948.

Strand, A. L.: "Strand and Spitzer Issue Statements on Spitzer's Dismissal," *Chemical and Engineering News* 27:906-9. 1949.

Wright, Sewall: "Dogma or Opportunism," *Bulletin of the Atomic Scientists* 5:141-42. 1949.

Zhdanov, V.: "Film on Ivan Michurin Coming to U.S.A.," *USSR Information Bulletin* 9:186-87. 1949.